Hubert Untersteiner

**Aquatic Invertebrates as Indicators to Pollution-induced Stress**

Hubert Untersteiner

# Aquatic Invertebrates as Indicators to Pollution-induced Stress

## Validation of the Locomotory Behaviour of Freshwater and Marine Crustaceans as Response to Sublethal Heavy Metal Stress with Video Imaging

Südwestdeutscher Verlag für Hochschulschriften

**Impressum/Imprint (nur für Deutschland/ only for Germany)**
Bibliografische Information der Deutschen Nationalbibliothek: Die Deutsche Nationalbibliothek verzeichnet diese Publikation in der Deutschen Nationalbibliografie; detaillierte bibliografische Daten sind im Internet über http://dnb.d-nb.de abrufbar.
Alle in diesem Buch genannten Marken und Produktnamen unterliegen warenzeichen-, marken- oder patentrechtlichem Schutz bzw. sind Warenzeichen oder eingetragene Warenzeichen der jeweiligen Inhaber. Die Wiedergabe von Marken, Produktnamen, Gebrauchsnamen, Handelsnamen, Warenbezeichnungen u.s.w. in diesem Werk berechtigt auch ohne besondere Kennzeichnung nicht zu der Annahme, dass solche Namen im Sinne der Warenzeichen- und Markenschutzgesetzgebung als frei zu betrachten wären und daher von jedermann benutzt werden dürften.

Verlag: Südwestdeutscher Verlag für Hochschulschriften Aktiengesellschaft & Co. KG
Dudweiler Landstr. 99, 66123 Saarbrücken, Deutschland
Telefon +49 681 37 20 271-1, Telefax +49 681 37 20 271-0, Email: info@svh-verlag.de
Zugl.: Graz, Karl-Franzens-Universität, Diss., 2004

Herstellung in Deutschland:
Schaltungsdienst Lange o.H.G., Berlin
Books on Demand GmbH, Norderstedt
Reha GmbH, Saarbrücken
Amazon Distribution GmbH, Leipzig
ISBN: 978-3-8381-0423-2

**Imprint (only for USA, GB)**
Bibliographic information published by the Deutsche Nationalbibliothek: The Deutsche Nationalbibliothek lists this publication in the Deutsche Nationalbibliografie; detailed bibliographic data are available in the Internet at http://dnb.d-nb.de.
Any brand names and product names mentioned in this book are subject to trademark, brand or patent protection and are trademarks or registered trademarks of their respective holders. The use of brand names, product names, common names, trade names, product descriptions etc. even without a particular marking in this works is in no way to be construed to mean that such names may be regarded as unrestricted in respect of trademark and brand protection legislation and could thus be used by anyone.

Publisher:
Südwestdeutscher Verlag für Hochschulschriften Aktiengesellschaft & Co. KG
Dudweiler Landstr. 99, 66123 Saarbrücken, Germany
Phone +49 681 37 20 271-1, Fax +49 681 37 20 271-0, Email: info@svh-verlag.de

Copyright © 2009 by the author and Südwestdeutscher Verlag für Hochschulschriften Aktiengesellschaft & Co. KG and licensors
All rights reserved. Saarbrücken 2009

Printed in the U.S.A.
Printed in the U.K. by (see last page)
ISBN: 978-3-8381-0423-2

DILUTION IS NO SOLUTION FOR POLLUTION

*« dosis facit venenum »*

Theophrastus Bombastus von Hohenheim
(Paracelsus) - 1493 - 1541

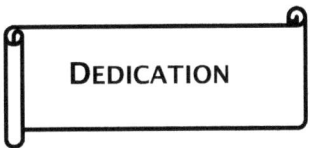

This work is dedicated to my family and to my mate Mag. Petra Schwarzl. Thank you for supporting and strengthen me to go my scientific way and thereby making this level of education possible.

## ACKNOWLEDGEMENTS

I would like to express my heartfelt gratitude to those who made the completion of this doctoral work possible.

First and foremost, I am very grateful to my supervisor and doctor father *Prof. Dr. Helmut Kaiser* for his patience, guidance, support and close friendship.

I am very thankful to *Ms. Jördis Kahapka* for joining our working group and doing her diploma thesis in the field of aquatic ecotoxicology.

I would like to thank the team of the *ARGE Umweltanalysen Ges.n.b.R. Graz* for the initiation and partially funding of this research project.

I am also very thankful to my mate *Mag. Petra Schwarzl* for reading the manuscript, improving my english and doing all this thousand little things in order to support me in my scientific work.

I am very thankful to *Mag. Gerwin Gretschel* (Director of the School of Marine Biology Valsaline – Pula/Croatia) and his team of marine biologists *Mag. Tom Puchner* and *Mag. Sonja Napetschnig* for supporting me during my ethotoxicological experiments with the marine crustacean *Hippolyte inermis* LEACH.

I also like to thank *Mr. Johannes Salvenmoser* for helping to sample the marine test organism *Hippolyte inermis* LEACH.

And last but not least I am very grateful to *Mag. Katja Pörtl* for reading the manuscript and for her expertly thoughts.

### THANK YOU ALL!

# TABLE OF CONTENTS

Dedication .................................................................................................................. i
Acknowledgement ..................................................................................................... ii
Table of contents ...................................................................................................... iii
Glossary .................................................................................................................... vi
Chapter I - General Introduction ............................................................................... 1
    I. 1. Heavy metals ................................................................................................... 4
        I. 1. 1. Cadmium (Cd) ........................................................................................ 5
            I. 1. 1. 1. Physical and chemical properties ..................................................... 5
            I. 1. 1. 2. Environmental behaviour and Ecotoxicology of Cd ........................ 5
            I. 1. 1. 3. Essentiality and toxicology of Cd .................................................... 6
        I. 1. 2. Copper (Cu) ............................................................................................. 7
            I. 1. 2. 1. Physical and chemical properties ..................................................... 7
            I. 1. 2. 2. Environmental behaviour and Ecotoxicology of Cu ........................ 7
            I. 1. 2. 3. Essentiality of Cu ............................................................................. 8
        I. 1. 3. Heavy metals in the marine environment ............................................... 9
    I. 2. Crustaceans as test organisms in aquatic biomonitoring ................................. 9
        I. 2. 1. Cladocerans ............................................................................................. 9
            I. 2. 1. 1. Taxonomy of the cladocera ............................................................ 10
            I. 2. 1. 2. Morphology and anatomy of the cladocera ................................... 10
            I. 2. 1. 3. *Daphnia magna* STRAUS ............................................................... 13
            I. 2. 1. 4. *Daphnia pulex* LEYDIG ................................................................ 13
        I. 2. 2. Decapoda ............................................................................................... 14
            I. 2. 2. 1. Taxonomy of the Hippolytidae ...................................................... 15
            I. 2. 2. 2. *Hippolyte inermis* Leach .............................................................. 15
    I. 3. Project aims .................................................................................................... 16
Chapter II – *Daphnia magna* STRAUS ...................................................................... 17
    II. 1. Introduction ................................................................................................... 18
    II. 2. Materials and methods .................................................................................. 20
        II. 2. 1. Testorganisms ...................................................................................... 20
            II. 2. 1. 1. Hatching of the ephippia ............................................................... 20
        II. 2. 2. Copper dilutions .................................................................................. 21

- II. 2. 3. Measuring of the swimming parameters .................................................... 21
- II. 2. 4. Statistics .................................................................................................... 23
- II. 3. Results ............................................................................................................... 23
- II. 4. Discussion .......................................................................................................... 26
- II. 5. Conclusions ........................................................................................................ 28

## Chapter III – *Daphnia pulex* LEYDIG .................................................................... 29

- III. 1. Introduction ....................................................................................................... 30
- III. 2. Materials and methods ...................................................................................... 30
  - III. 2. 1. Test organisms ............................................................................................ 30
    - III. 2. 1. 1. Hatching of the daphnids ................................................................... 30
  - III. 2. 2. Heavy metal dilutions ................................................................................. 31
  - III. 2. 3. Measuring of the locomotory activity ........................................................ 31
- III. 3. Results ............................................................................................................... 31
- III. 4. Discussion .......................................................................................................... 34
- III. 5. Conclusion ......................................................................................................... 34

## Chapter IV – *Hippolyte inermis* LEACH .................................................................. 35

- IV. 1. Introduction ....................................................................................................... 36
- IV. 2. Material and methods ....................................................................................... 37
  - IV. 2. 1. Study sites and test organisms ................................................................... 37
  - IV. 2. 2. Cadmium dilutions ..................................................................................... 38
  - IV. 2. 3. Measuring of the locomotory activity ........................................................ 39
  - IV. 2. 4. Statistics ..................................................................................................... 40
- IV. 3. Results ............................................................................................................... 40
- IV. 4. Discussion .......................................................................................................... 43

## Chapter V – General Discussion ................................................................................ 46

- V. 1. Video analysis ..................................................................................................... 46
  - V. 1. 1. Analysis software ........................................................................................ 46
- V. 2. Copper effects on daphnids ................................................................................ 47
- V. 3. Copper effects on other aquatic invertebrates .................................................... 48
  - V. 3. 1. Acute toxicity of Cu .................................................................................... 48
  - V. 3. 2. Physiological and sublethal effects of Cu to aquatic invertebrates ............. 49
    - V. 3. 2. 1. Bioconcentration of Cu ....................................................................... 50

## Table of contents

- V. 4. Cadmium effects on crustaceans ................................................................. 52
  - V. 4. 1. Acute toxicity .......................................................................................... 52
  - V. 4. 2. Physiology and sublethal effects of Cd to aquatic invertebrates ............. 52
    - V. 4. 2. 1. Bioconcentration of Cd ..................................................................... 53
- V. 5. Mechanisms of metal detoxification in organisms ......................................... 54
  - V. 5. 1. Metallothioneins (MTs) ........................................................................... 54
- V. 6. Biomonitoring of xenobiotica ...................................................................... 55
- VI. References ..................................................................................................... 57
- VII. Attachments .................................................................................................. 75
- VIII. Summary .................................................................................................... 106

# GLOSSARY

**Bioconcentrationfactor: (BCF):** Relation between the concentration of a chemical compound in the organism in comparison to the concentration in the environment.

**Bioaccumulation:** accumulation of a chemical compound in the tissues of an organism.

**Bioavailability:** a chemical compound is bioavailable, when it is in a structure which can be taken up by organisms.

**Bioassay:** standardised biotest using standardised test organisms.

**Bioindicator:** species which indicate a certain situation in the field.

**Biological end point:** the response of an organism which is measured in a toxicity test; choice of end point depends upon the purpose of the test.

**Biomagnification:** accumulation of a chemical compound over the food web.

**Biomarker:** a biochemical, cellular, physiological, morphological or behavioural response at the (sub)organism level to sublethal exposure to chemicals.

**Chemical Abstract Service (CAS Number):** Unique Code for each chemical species.

**Cytochrome $P_{450}$:** a large group (>70, in 14 families) of iron-containing (haem) proteins that catalyse many biological oxidations, including the metabolism of a wide variety of xenobiotics.

**DIN:** Deutsches Institut für Normung (German Institute for standardization)

**$EC_{10}$:** effective concentration of a xenobiotic causing an effect in 10 % of test organisms.

**$EC_{50}$:** median effective concentration: concentration of a xenobiotic causing a designated effect other than mortality in 50% of test organisms.

**Filial generation ($F_{1,2}$):** Filialgeneration = Generation of descendants.

**Hazard:** potential of a chemical to cause harm.

**ISO:** International Organisation for Standardization.

**$LC_{10}$:** lethal concentration, which causes a 10 % mortality of the test organisms.

**$LC_{50}$:** median lethal concentration: concentration which kills 50% of test organisms.

**Metallothionein:** metal-binding protein; can be important in detoxification processes.

**NOEC:** no observed effect concentration.

**OECD:** Organization for Economic Co-operation and Development

**ÖNORM:** Österreichisches Normungsinstitut (Austrian Standards Institute)

**ppb:** parts per billion: e. g. [$\mu g\ l^{-1}$]

**ppm:** parts per million: e. g. [$mg\ l^{-1}$]

**stressor:** any physical, chemical or biological entity or process that can induce adverse effects on individuals, populations, communities and ecosystems.

**test species:** is a standardised species which is used in biotests.

# Chapter I – General Introduction

The biota of an ecosystem is influenced by a variety of biotic and abiotic stress factors. Especially with the onset of modern industrialization our environment is increasingly burdened with pollutants, as for instance heavy metals (e. g. mercury, cadmium, lead, copper), semi-metals (e. g. antimony, arsenic), organo-metals and the great spectrum of organic pollutants (e. g. PCB, PAK) (Oehlmann and Markert, 1999).

Today more than 5 million chemical compounds are known and about 80 000 of them are in use (Fent, 1998). Also, many new chemicals are being developed annually for a wide variety of agricultural, industrial and other applications (Cleveland et al., 1999). Being aware of this fact it is obvious that one of the main environmental problems society is facing today is the burden of the environment with pollutants with all negative consequences to complex ecosystems as for instance the loss of biodiversity and natural resources.

The necessity of the establishment of an efficient detection system for toxic water substances in Austria was recognized due to the great die-off of the fish-population in the river Danube in the beginning of 1970 (Rodinger, 1994).

However, the major incident which revived the concern about the water quality was a calamity in 1986, when the runoff from a fire in the Sandoz plant at Basel (Switzerland) wiped out part of the biota in the river "Rhine" (Hendriks and Stouten, 1993). The new gained higher consciousness for environmental protection stimulated an international cooperation within Europe which led to the setup of the "Rhine Action Program" (APR) with the aim of reduction of pollution and the restoration of destroyed habitats, thus enabling the resettlement of indigenous species (e. g. *Salmo salar*) (UNCED, 1992; WIR, 1995).

In principle the conventional approach to control harmful chemicals in aquatic ecosystems is to use a set of global physical-chemical parameters.

General Introduction

Because of the fact, that water pollution is a complex situation chemical procedures alone can not provide sufficient information on the potential harmful effects of chemicals in the aquatic environment (Slabbert and Venter, 1999).

Due to this fact biological assays have become important tools to assess harmful chemical activity and to assist in developing precautionary measures and strategies for environmental management.

There are some points which highlight the advantages of biomonitoring systems (Gerhardt, 1995):

- The entire mixture of toxicants will be taken into account, including synergistic and antagonistic effects.

- Biomonitoring puts the organism in focus instead of giving concentration levels of toxic substances which do only in a limit way tell anything about the bioavailability of the compounds on different levels of biological organisation (organism-, population- or ecosystem level).

- Moreover Biomonitoring allows a temporal integration of the effects of different pollutant exposures throughout the life cycle of an organism.

In Europe and in the United States of America numerous bioanalytical techniques for the control of water quality have been developed and applied at the sub- and multicellular levels of biological organization.

Tab. 1 gives a short overview about bioassays and active and passive biomonitoring systems developed and used during the last decades in Europe and in the US.

General Introduction

**Tab. 1:** Overview about developed and applied biological assays (completed and modified after Gerhardt, 1999)

| Biological organization | Endpoint of biological response | Reference(s) |
|---|---|---|
| **Biomarker** | | |
| stress proteins (e. g. Hsp27, Hsp60), metabolic enzymes (e.g. hexokinase, dehydrogenase), | Induction | Schramm, M. et al. (1999) |
| biotransformation enzymes (cytochrome P450 enzyme system resp. mixed-function oxygenase system) | catalytic activity | |
| **Fish cell lines** | | |
| cells of connective tissue (e. g. R1-cells, RTG-2-cells) Hepatocytes (PLHC-1, RTL-W 1-cells) | acute cell death cell vitality morphological alterations biochemical alterations | Braunbeck et al., 1992 Braunbeck, 1995 |
| **Bacteria** | | |
| *Vibrio fisheri* (NRRL-B-11177) | bioluminescence | Krebs, 1992; Link, 1992; Steinhäuser, 1992; Klein, 1992; Bulich et al., 1996 |
| *Synechococcus sp.* (PCC6301) | electron transport | Stein, 1992 |
| *Escherichia coli* | respiration | Stein, 1992 |
| *Salmonella typhimurium* | genotoxicity | Nakamura et al., 1987 |
| **Algae** | | |
| *Chlamydomonas reinhartii.* | growth inhibition $O_2$-production fluorescence | Gerhardt., Putzger, 1992; Merschhemke, Regh,1992 |
| *Selenastrum capricornutum* | growth inhibition | Rodinger, 1997; Chiaudani, Vighi, 1978 |
| *Euglena gracilis* | motility | Stallwitz, Häder, 1994 |
| **Higher plants** | | Eberius, Vandenhirtz, 1999 |
| *Vicia faber* | $O_2$-production | Overmeyer et al., 1994 |
| *Lepidium sativum* | germination root length | Neururer, H., 1975 |
| *Lemna minor* | growth inhibition, vitality | |
| **Mussels** | | |
| *Dreissena polymorpha* | shell movements | Bocherding, Volpers, 1994; Hoffmann et al., 1994 |
| *Corbicula fluminea* | shell movements | Ham, Peterson, 1994 |
| *Anodonta cygnea* | shell movements | Englund et al., 1994 |
| *Mytilus galloprovincialis* | metallothionein concentration in tissue, metal concentration in tissue ; morphological alterations | Viarengo et al., 1997; Rainbow et al., 2000 ; Fichet, Miramand, 1998 |
| *Adamussium colbecki* | | |
| *Mytilus trossulus* | | |
| *Crassostrea gigas* | | |
| **Crustacea** | | |
| *Daphnia spp.* | motility, swimming velocity, swimming direction, phototaxis, trace metal concentrations in tissue | Knie, 1978; Baillieul, Scheunders, 1998; Blübaum-Gronau, Hoffmann, 1998; Fichet et al., 1998; Rainbow et al., 2000, this work |
| *Artemia franciscana* | | |
| *Gammarus* spp. | | |
| *Balanus improvisus* | | |
| *Hippolyte inermis* | | |
| **Echinoderms** | | |
| *Paracentrotus lividus* | bioaccumulation, morphological alterations during development | Fichet et al., 1998 |

**Tab. 1:** Continuation

| Biological organization | Endpoint of biological response | Reference(s) |
|---|---|---|
| **Polychaetes** | | |
| *Arenicola marina* | mortality and volume regulation | Rasmussen, Andersen, 2000 |
| **Craniota** | | |
| *different species of fish* | rheotaxis, motility ventilation, heart rate, locomotion | Spieser et al., 1994 |
| *Fundulus heteroclitus* | Mortality | Hurk, et al., 1998 |
| *Gnathonemus petersii* | Electric Organ Discharge (EOD) | Thomas et al., 1996 |
| *Apteronotus albifrons* | | |
| **Ecosystem** | Species spectrum | Schwoerbel, 1994 |
| saprobic classification | (limnosaprobical organisms) | |

## I. 1. Heavy metals

Heavy metals are defined as metals with a density $> 6$ g cm$^{-3}$ (Gunkel, 1994; Fent, 1998). These metals on the one hand include essential metals like iron, copper and zinc and on the other hand non-essential metals like cadmium, lead and mercury (Fent, 1998). Although essential metals (including their salts) are very important for a variety of physiological processes in most organisms they can be in the same way toxically to organisms like non-essential metals, if the concentration exceeds a specific level.

The critical concentration level depends on the metal species, on the physical-chemical environmental conditions and on the biological system (Koch, 1995). The inherent toxicity of a metal depends upon its capacity to disturb the dynamic life processes in biological system by combining with cell organelles, macromolecules and metabolites.

In this work the toxic effects of two important heavy metals, namely the essential copper (Cu) and the non-essential cadmium (Cd), to different crustacean species (*Daphnia magna*, *Daphnia pulex* and *Hippolyte inermis*) have been investigated.

# General Introduction

## I. 1. 1. Cadmium (Cd) – CAS-Number: 7440-43-9

### I. 1. 1. 1. Physical and chemical properties

Cadmium is a silver-white, shiny and expansive heavy metal with a relative atom weight of 112.40. It is the second element in the group II b of the periodic table of elements. There exist 8 stable Cd-Isotopes and 20 unstable Cd-Isotopes. The proportion of Cadmium in the earth's crust amount to $3 \times 10^{-5}$ weight-percent (Breuer, 2000).

Cd is one of the most toxic metal. Acute and chronically intoxication of humans at vocational exposure have been reported (Koch, 1995). Cadmium intoxication to humans has firstly been reported in Japan in 1947 (Itai-itai-disease) (Nentwig, 1995).

### I. 1. 1. 2. Environmental behaviour and Ecotoxicology of Cadmium

Cd is a trace metal and is integrated in a permanent biogeochemical cycle. The naturally input into the environment amounts ~ 800 tons per year. In comparison to anthropogenic emissions the naturally input is relatively low. Tab. 2 gives an overview about the increasing Cd-concentration in the environment over the time.

Tab. 2: Overview about the increasing Cd-concentration in the environment over time (modified after Koch (1995)).

| compartment | Year | | | |
|---|---|---|---|---|
| | 1750 | 1930 | 1970 | 2100 |
| air [ng m$^{-3}$] | 0 | 2.4 | 5.7 | 8.7 |
| soil [ppb] | 100 | 160 | 320 | 750 |
| river sediments [ppb] | 150 | 900 | 3900 | 15000 |
| river water [ppb] | 0.03 | 0.18 | 0.78 | 3.0 |

As already mentioned most of the cadmium in the environment comes from man made sources, especially from non-ferrous smelting, fuel combustion, disposal of cadmium-containing products and application of phosphate fertilizer. Other contributing sources include metal and phosphate fertilizer production and combustion of fossil fuels, wood and waste incineration.

General Introduction

The toxicological potential of Cd to aquatic Organism depends on the metal species, the physical-chemical characteristics of the water (e. g. pH, salinity, calcium- and oxygen concentration) and on the specific physiology of the organisms. Cadmium gets enriched in the soil or sediment, respectively. The retention time in the hydrosphere is assumed to be about 2 years.

Cadmium is usually found in the form of cadmium oxide, chloride, sulphate or sulphide. In air, cadmium exists as fine particulate. The main chemical species in the atmosphere is cadmium oxide although other cadmium salts may also be present. The cadmium compounds in air is stable and undergoes little atmospheric transformation. Cadmium particulate is dispersed by wind and eventually either settled out by rain or snow or dry deposited on land or surface water. The fine particulate containing cadmium can remain airborne for days to weeks and travel hundreds to thousands of kilometres (OECD, 1994).

In water, cadmium exists either dissolved or as part of insoluble complexes. Solubility is promoted by acidic conditions. Soluble cadmium is quite mobile in water and in soil. Partitioning into the sediment is enhanced by precipitation and sorption to mineral surfaces and organic materials as well as by action of sediment bacteria. The levels in the sediment tend to be at least an order of magnitude higher than in the overlying water column. High levels of organic material in the water promote formation of organic complexes with cadmium, which are poorly soluble. Also in reducing environment, cadmium may precipitate out as cadmium sulphide. Cadmium in soils may leach into water, especially under acidic conditions. Transformation processes for cadmium in soil are mediated by sorption from and adsorption to water, and include precipitation, dissolution, complexation, and ion exchange (WHO, 1992).

I. 1. 1. 3. Essentiality and toxicology of Cd on aquatic animals

Cd is considered not to be essential for organisms. The acute toxicity of Cd to pisces ranges from 23 to 140 ppm. The acute toxicity of crustaceans ranges from 1 to 10 ppm. Aquatic vertebrates seem to be less sensitive to Cd-stress than aquatic invertebrates. Pisces tolerate a Cd-concentration in a range from 1 to 10 mg $l^{-1}$ without toxically damages. Very important for the biota is the fact, that Cd can be accumulated in the tissue of the organisms (bioaccumulation).

General Introduction

Cadmium can be taken up and retained by aquatic and terrestrial plants and can substantially bio concentrated in aquatic invertebrates and fish. In terrestrial animals, cadmium is particularly concentrated in the liver and kidney of animals that eat the plants. Low soil pH tends to increase the availability of cadmium.

I. 1. 2. Copper (Cu) – CAS-Number: 7440-50-8

I. 1. 2. 1. Physical and chemical properties

Copper is a brightly red, relatively soft, tough and expansive heavy metal with a relative atom weight of 63.55. It is the first element of group 1 b of the periodic table. There exist 2 stable Cu-Isotopes and 12 unstable Cu-Isotopes. The proportion of Copper in the earth's crust amount to $5 \times 10^{-3}$ weight-percent (Breuer, 2000).

I. 1. 2. 2. Environmental behaviour and Ecotoxicology of Copper

Cu is widely distributed in different environmental media because it is a naturally occurring element. Natural discharges to air and water may therefore be significant. Tab. 3 gives an overview about worldwide emissions of Cu into the environment at the beginning of the 1980s.

Tab. 3: Overview about the worldwide emissions of Cu into the environment at the beginning of the 1980s (modified after Pacyna et al. 1995).

| compartment | type | Subtotal [$10^3$ t yr$^{-1}$] | Total [$10^3$ t yr$^{-1}$] |
|---|---|---|---|
| Air | natural | 2.2 – 53.8 | 21.9 – 104.6 |
|  | anthropogenic | 19.7 – 50.8 |  |
| Soil | anthropogenic |  | 541.5 – 1402.8 |
| Water | anthropogenic |  | 34.7 – 190.5 |

Due to the fact that Cu is a naturally occurring element it is widely distributed in water, whereby the copper levels in freshwater (mainly rivers) range from 0.6 to 400 ppb [µg l$^{-1}$] (Alberta Water Guideline, 1996).

General Introduction

In 1985 limnological investigations of the styrian government showed that the river Mur (Styria, Austria) was intensively burdened by Cu. In some sample sites top values of Cu concentrations up to 120 ppb have been measured (Pescheck and Herlicska, 1990). In the meantime the water quality of the Mur is satisfying due to strict environmental protection laws.

Cu displays for oxidation states: Cu, $Cu^+$, $Cu^{2+}$, $Cu^{3+}$. Among these ions $Cu^{2+}$ is the most important oxidation state and is the oxidation state generally encountered in water. In freshwater free of organic complexing agents, the solubility of $Cu^{2+}$ is controlled by malachite ($Cu_2(OH)_2CO_3$) below a pH of 7 and by tenorite (CuO) above a pH of 7 (Stumm, Morgan, 1981). The main inorganic cupric ion species present in freshwater vary with pH, the dominant species change from $Cu^{2+}$, $CuCO_3$, $Cu(CO_3)_2^{2-}$, $Cu(OH)_3^-$ to ultimately $Cu(OH)_4^{2-}$ (Stumm, Morgan, 1981). Further parameters, which influence the relative proportion of the cupric ion species, are the alkalinity and the magnitude of stability constants for the formation of complexes. Although pH and alkalinity vary substantially in freshwater systems, $Cu^{2+}$, $Cu(OH)^+$, $Cu(OH)_2$ and $Cu(CO_3)_2^{2-}$ make up to 98 % of dissolved inorganic copper (Nelson et al., 1986). In water copper is held in solution mainly by complexation with naturally occurring organic ligands. In freshwater, organic ligands are more important in binding copper than inorganic ligands (Spear and Pierce, 1979).

Cupric ion is the main toxic species. However, Cu species other than $Cu^{2+}$ are toxic to daphnids (Dave, 1984; Borgmann and Charlton, 1984).

I. 1. 2. 3. Essentiality of Cu on aquatic animals

In general Cu is considered to be an essential compound for most living organisms (plants and animals). It is a component of a variety of metalloenzymes and respiratory pigments (Demayo and Taylor, 1981; Alberta Water Quality Guideline, 1996). In plants it is required for the synthesis of chlorophyll (photosynthetic pigment) and in animals it is required for the synthesis of haemoglobin (respiratory blood pigment). Cu serves as the oxygen coupling site in haemocyanin, the respiratory blood pigment in many molluscs and certain other invertebrates, like crustaceans (e. g. gammarids). However, copper becomes toxic to aquatic biota, when biological requirements are exceeded.

## I. 1. 3. Heavy metals in the marine environment

Heavy metal pollution is not only a problem in freshwater ecosystems, but also plays a major role in the marine environment, which has been discussed by several studies (Tankere and Statham, 1996; Sekulić and Vertačnik, 1997; Zago et al., 2000; Kljaković Gašpic et al., 2002).

Marine ecosystems are strongly influenced by freshwater discharges of the rivers (e. g. the river Po influences mainly the northern part of the Adriatic Sea). The toxic potential of heavy metals in marine ecosystems is to interpret as very high, since continuous exposure of marine organisms to heavy metal concentrations may result in bioaccumulation, and subsequent transfer to man through the food web.

## I. 2. Crustaceans as test organisms in aquatic biomonitoring

Within the framework of their evolution, crustaceans became a very successful group of animals. They distributed in a number of different aquatic and terrestrial habitats. Due to this fact crustaceans are frequently used organisms in the context of classical biomonitoring (Rinderhagen et al., 1999). A further point that makes crustaceans very practicable for standardised biomonitoring is the fact, that many families of this group are easy to hold and rear under laboratory (artificial) conditions. Many of characteristic features, like reproduction strategies, morphological, behavioural and physiological parameters, are used as biological endpoints in aquatic biomonitoring.

### I. 2. 1. Cladocerans

Cladocerans, or water fleas, are small (0.2-6 mm) aquatic crustaceans forming one of the classes of the Branchiopoda. Most inhabit quiet fresh waters (seas, ponds). Along with the rotifers and copepods they account for most of the freshwater zooplankton yet most of them are benthic. Most cladocerans are filter feeders that consume phytoplankton, which they remove from the water using their setose thoracic appendages. A few are carnivores preying on other cladocerans.

General Introduction

Since the early beginning of the development of biomonitoring techniques, cladocerans, like the dahpnids (e. g. *Daphnia magna*, *Daphnia pulex* and other Genera) have been recognized as very practicable and suitable indicator organisms for xenobiotica pollution in aquatic ecosystems (Fomin et al, 2003).

In the meantime there exist a lot of standardized biotests (after national and international norms like ÖNORM, DIN, OECD and ISO) using these organisms as test organisms.

I. 2. 1. 1. Taxonomy of the Cladocera

Recent molecular evidence suggests that the Cladocera is a monophyletic group (Crease and Taylor, 1998; Hanner, 1997), contrary to previous classification schemes, which divided the Cladocera into four orders (e.g., Dodson and Frey, 1991). The order Cladocera is now divided into four suborders, 11 families, about 80 genera and roughly 400 species, although taxonomic revision in all families is underway so that the numbers of genera and species are likely to increase.

At present, only the *Daphnia* section is complete:

|  |  |
|---|---|
| Phylum: | Arthropoda LATREILLE, 1829 |
| Subphylum: | Crustacea LAMARCK |
| Class: | Branchiopoda LATREILLE, 1817 |
| Subclass: | Diplostraca GERSTAECKER, 1866 |
| Order: | Cladocera LATREILLE, 1829 |
| Suborder: | Anomopoda |
| Family: | Daphniidae STRAUS, 1820 |
| Genera: | *Daphnia* MÜLLER, 1785 |

I. 2. 1. 2. Morphology and Anatomy of the Cladocera

In most cladocerans the body is almost entirely enclosed in a large bivalved carapace from which the head extends anteriorly. The mouth is located on the ventral head and points posterioly. The body is laterally compressed.

The enlarged second antennae serve as locomotory organs and the first antennae are vestigial in females and not much larger in males. Many cladocerans, including *D. magna* and *D. pulex*, have a single, tiny, median, naupliar eye or ocellus posterior-ventral to the much larger compound or complex eye. It is embedded in the edge of the brain.

The thoracic region makes up most of the body and is entirely enclosed in the carapace. It is immediately posterior to the head and bears 5-6 pairs of biramous, setose thoracic appendages, the distal tips of which may extend from the gape of the carapace. These appendages and their setae are used as filters to remove phytoplankton from the water. Once strained from the water the cells are passed anteriorly from appendage to appendage until they reach the mouth.

The oval heart is a conspicuous feature of the dorsal region of the anterior thorax. It bears a single pair of ostia. Contractions of the heart force blood anteriorly into the head from which it flows posteriorly into the thorax via three channels. The two lateral channels each serve one side of the carapace whereas the median channel runs ventral to the gut and gives off branches to the thoracic appendages. Blood returns to the heart from each of these areas.

Posteriorly the thorax narrows to become the postabdomen (or abdomen). This region lacks appendages but is flexible and highly mobile. It is usually bent so that it is tucked underneath the thorax, where it is completely enclosed in the carapace. The posterior tip bears a pair of postabdominal claws and the anus opens on the dorsal surface of the postabdomen. There is a pair of long setae on the dorsal edge of the postabdomen. The postabdomen can be straightened so that it extends posteriorly from the carapace. The postabdomen and its claws are used to clean the thoracic appendages.

Most species of *Daphnia* possess a posterior apical spine on the carapace that may make its owner more difficult to consume by predators. Many species also have a spine on the head. Planktonic cladocerans are preys from zooplanktivorous fishes and invertebrates.

## General Introduction

The C-shaped intestine extends from the mouth, through the dorsal thorax and through the postabdomen to the anus near the distal end of the postabdomen. Food gathered by the thoracic appendages moves anteriorly along a ventral food groove, propelled by the same thoracopod movements that filter the food from the water in the first place. Upon reaching the anterior end of the thorax the food moves into the posteriorly directed mouth.

The gonads are long tubes or sacs extending most of the length of the thorax and lying on either side of the intestine. The gonoducts open to the exterior via gonopores posterior to the last pair of thoracic appendages. In females the ovary opens dorsally, via an oviduct, into the brood chamber. In males the vas deferens leads to a ventral gonopore.

Female cladocerans have a large open brood chamber, located under the dorsal carapace in the posterior thoracic region. In this brood chamber eggs are deposited and brooded during they complete embryonic development.

Females produce two types of eggs. Summer eggs have little yolk and develop parthenogenetically, without fertilization. Summer eggs are carried in the brood chamber at least until they hatch and in some species until they are sexually mature and have young of their own.

Winter eggs, on the other hand, are very yolky and are produced only after fertilization. These eggs are also released into the brood chamber, where they are fertilized. Winter eggs, unlike those of summer, are not brooded rather are immediately released by the female, either enclosed in a protective cuticular ephippium or naked. The ephippium is released when the female molts. Ephippia may sink or float, depending on species. Winter eggs hatch in the following spring. Winter eggs always hatch into parthenogenetic females, i.e. females that reproduce without fertilization. Eventually, after one or several generations of parthenogenic females and their summer eggs, males will be produced and fertilization will occur to produce a new generation of over wintering eggs (Fox, 2003).

General Introduction

An overview about the morphology and anatomy of a daphnid is shown in Fig. 1.

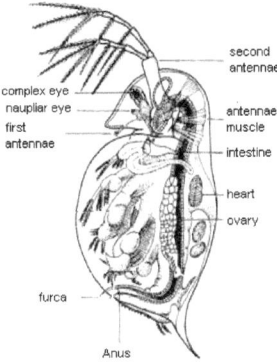

**Fig. 1:** Schematic habitus of a *Daphnia* species, lateral view (modified from Anonymus, 2003).

I. 2. 1. 3. *Daphnia magna* STRAUS

*D. magna* STRAUS is one of the largest daphniids with adult females ranging in size from 2.0 to 5.0 mm. This species is widely distributed in most habitats of the Holarctic. *D. magna* is most common in intermittent ponds, especially those which are either eutrophic or slightly saline (up to 2000 microsiemens/cm). Males and females are easily distinguished from other daphniids, because of their strongly incised postabdomen. *D. magna* reproduces by cyclic parthenogenesis and is not known to hybridise with any other species.

I. 2. 1. 4. *Daphnia pulex* LEYDIG

*D. pulex* LEYDIG shows much variation in body size with mature females ranging in length from 1.1 to 3.5 mm. The species is common in ponds throughout most of temperate Holarctic, but is absent from the subtropics and high arctic. As the most common member of the *pulex* group, this species has been confused with each of the other 15 members of this group. Its lack of elongate setae on the ventral carapace and of cuticular melanization coupled with its pubescent abdominal processes and dense spinescence on the dorsal surface of its ephippium distinguish *D. pulex* from all of these species excepting *D. pulicaria*.

- 13 -

These two species show some divergence in habitat use, with *D. pulicaria* absent from intermittent ponds, while *D. pulex* never occurs in lakes.

However, the species do often co-occur in permanent ponds. Discrimination of the species is further complicated by the prevalence of $F_1$-hybrids between these species. As a result, definitive taxonomic assignments are reliant upon allozyme analysis. *D. pulex* shows variation in its breeding system with populations reproducing by either cyclic or obligate parthenogenesis.

I. 2. 2. Decapoda

The Decapoda includes ~10 000 species and thus it is the most biodiverse order of crustaceans. Decapods are distributed over all oceans. They are mainly benthic organisms and are distributed from the seashore to the deep sea. Most species of decapods inhabit the littoral zone. Although shrimps are good swimmers, decapods are hardly represented in the pelagic zone.

In comparison to the cladocerans, decapods play a minor role in classical biomonitoring. However, some species of the decapods seem to be very reliable indicators to xenobiotica stress. Brouwer and Brouwer (1998) for example investigated biochemical defense mechanisms against Cu-induced oxidative damage in the blue crab (*Callinectes sapidus*). Crayfish of the species *Cambarus bartoni* were used for studying the bioconcentration of Cu to the different tissues of the organisms (Alikhan et al., 1990).

In this work the species *Hippolyte inermis* LEACH have been used for studying acute and chronic Cd-effects.

## I. 2. 2. 1. Taxonomy of the Hippolytidae

Recently there are described 15 groups within the family of Hippolytidae.

| | |
|---|---|
| Phylum: | Arthropoda LATREILLE, 1829 |
| Subphylum: | Crustacea LAMARCK |
| Class: | Malacostraca LATREILLE, 1806 |
| Order: | Decapoda |
| Suborder: | Pleocyemata |
| Family: | Hippolytidae |
| Genus: | *Hippolyte* |

## I. 2. 2. 2. *Hippolyte inermis* LEACH

The species *Hippolyte inermis* LEACH is characterised through a short, 3-segmental carpus of the $2^{nd}$ pereiopod. The rostrum is very long and exceeds the length of the cephalothorax. The seagrass shrimps are mostly green coloured, but also rarely brown (Riedl, 1983).

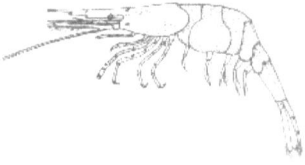

Fig. 2: Schematic habitus of *Hippolyte inermis* LEACH, lateral view (modified from Riedl, 1983).

The seagrass shrimp is an inhabitant of seagrass meadow habitats covered with different species of seagrass (e. g. *Posidonia* sp., *Cymodocea* sp. or *Zostera* spp.). In these habitats the shrimps are well camouflaged.

## 1. 3. Project aim(s)

The aim of this work is to analyse the behavioural response of different aquatic invertebrate species, like the freshwater crustaceans *Daphnia magna* STRAUS and *Daphnia pulex* LEYDIG and the marine crustacean *Hippolyte inermis* LEACH to heavy metal stress and to determine reliable biological endpoints by means of video technique. Reliable behavioural parameters, which can be used for evaluating the toxic potential of heavy metals, have to be at least as sensitive as the biological endpoint "mortality", which is commonly used in the determination of $LC_{50}$ values.

The used daphnids (*D. magna* and *D. pulex*) are already standardised and commonly used organisms in aquatic ecotoxicology. However, the marine crustacean *H. inermis* has not been used as test organism in the context of classical biomonitoring before. Thus, a further point of interest of this study is the question, whether this organism can be used as standardised test organism.

The paramount aim is the advancement of the use of biomonitor systems as a supplement for physical-chemical surveillance of water quality in freshwater and marine ecosystems. Doing so, this study follows the important demands of the Agenda 21 (Bundesministerium für Umwelt, Naturschutz und Reaktorsicherheit, 1997) and the EU Water Framework Directive, which emphasize the protection of freshwater as a restricted resource (Gerhardt, 1999; UNCED, 1992).

# Chapter II – *Daphnia magna* STRAUS

**Behavioural response of the cladoceran *Daphnia magna* STRAUS to sublethal Copper stress - Validation by image analysis**[1]

with Jördis Kahapka

## Abstract

In this study changes in the locomotory behaviour of the freshwater cladoceran *Daphnia magna* STRAUS were used as sublethal indicators of toxic copper (Cu) stress. The behavioural responses were determined by a real time image analysis, using a video camera and a Pentium-PC equipped with a standard low cost frame grabber. For a sequence of 50 images per group, where 10 daphnids were moving simultaneously, the trajectories have been reconstructed in binary image sequences. As biological end points we defined the average swimming velocity and the average duration of swimming activity and inactivity. The behavioural responses of the daphnids were analysed under normal conditions (without Cu stress) and after application of sublethal Cu stress of following concentrations [in µg l$^{-1}$ (ppb)]: $C_1 = 1$, $C_2 = 5$, $C_3 = 10$, $C_4 = 20$, $C_5 = 30$. The test organisms were exposed to the Cu concentration for 24 hours under static conditions. Already after 9 hours of Cu-exposure a significant (* $P < 0.05$) decrease of the average swimming velocity could be observed at the group of the highest Cu concentration (30 ppb). After 13 hours of Cu-contamination the swimming velocity was significantly (* $P < 0.05$) reduced at the group of 20 ppb Cu-treatment and after 14 hours a significant (* $P < 0.05$) decrease of the average swimming velocity could be measured at the group of 10 ppb Cu-treatment. No significant decrease of the swimming velocity could be observed in the 1 ppb and 5 ppb Cu-treatment.

*Key words:* Ethotoxicology; Biomonitoring; Copper; Daphnids; Sublethal stress; Swimming behaviour

---

[1] published in Aquatic Toxicology 65 (2003), 435 – 442.

## II. 1 Introduction

Cladocerans are ecologically very important members of freshwater invertebrates and amongst them *Daphnia magna* STRAUS has been often utilized as test organism for the ecotoxicological monitoring of aquatic ecosystems. In the evaluation of the toxically potential of xenobiotics behavioural parameters are accurate and reliable indicators, since the behaviour of an organism is the end point of a sequence of neurophysiological events including stimulation of sensory and motor neurons, muscular contractions and release of chemical messages (Lagadic et al., 1994; Gerhardt, 1995). Especially monitoring of the locomotory behaviour plays an important role in the evaluation of the burden of an ecosystem with toxic compounds. Based on behavioural responses to xenobiotica stress, a variety of automated biological sensor systems (online biomonitoring) have been developed during the last decades using different aquatic invertebrates and vertebrates as sensor organisms (e. g. Fish-Rheotaxis-test, Koblenz behavioural fish test, Dreissena-Monitor, Mussel-Monitor and the Dynamic Daphniatest) (Knie, 1978; Bocherding, 1992; Schmitz et al., 1994; Gerhardt, 1999). It is known that for permanently swimming organism like the freshwater cladoceran *D. magna* STRAUS the swimming activity is closely connected to its energy metabolism and external environmental conditions (Baillieul and Blust, 1999). Changing external conditions like the burden of its habitat with toxic compounds or the appearance of predators induces a stress situation by disturbing the normal functions. Thus the daphnids are forced to use a part of the metabolism energy to restore this imbalance (stress response). *D. magna* STRAUS reacts to such stress-situations by escape -, adaptation - or protection reactions (Wolf et al., 1998). The escape reaction is characterised by an increased swimming activity. During adaptation reaction a part of the energy will be used for adaptional mechanisms and the protection reaction is characterised by a decreased swimming activity due to the loss of coordination (Ferrando and Andreu, 1993; Wolf et al., 1998).

In this paper the behavioural responses of *D. magna* STRAUS to sublethal concentrations of Cu were investigated. The main objective was to evaluate this ethotoxicological biomonitoring system measuring swimming parameters of simultaneous swimming daphnids using copper (Cu) as stress factor.

We hypothesised that a decreased fitness of the daphnids, evaluated in a first line in form of the behavioural parameter "swimming velocity" and in a second line in form of the parameter "swimming participation" is a function of chronic Cu contamination. A further point of interest concerning the sensitivity of the selected behavioural endpoints was the question, after what time span the chronic copper stress can be detected. The heavy metal copper was chosen, because of its physiological and toxicological features. It is an essential micronutrient for most living organisms, being incorporated in at least 30 different enzymes like haemocyanin, which is responsible for oxygen binding and oxygen transport in crustaceans (Gerhardt 1995; Herkovits and Helguero, 1998). However, it is known that *Daphnia spp.* have extracellular haemoglobin instead of haemocyanin as respiratory pigment in their body fluid (hemolymph) (Dave, 1984; Sell, 1998; Pirow et al., 2001) and although the functions and benefits of haemoglobin in *Daphnia* species are already well investigated (Sell, 1998; Wiggings and Frappell, 2000; Pirow et al. 2001; Zeis et al., 2003) only less literature is available which deals with the concentration-dependent influence of the heavy metal Cu on the haemoglobin content of *D. magna* in the context of classical biomonitoring. Dave (1984) showed in his study three different effects of Cu on the haemoglobin content in *D. magna*: (1) a decrease between 0.001 and 0.05 µg Cu $l^{-1}$, (2) an increase up to 1.6 µg Cu $l^{-1}$, and (3) a sudden increase at higher concentrations. This complex pattern of a dose-response relationship between Cu and haemoglobin content may be lead back to Cu effects on metabolic processes which regulate the internal oxygen concentration in exposed animals. Dave (1984) suggested that Cu might act through the same mechanism as for dissolved oxygen in the water.

The ecotoxicological importance of Cu is in so far given as it is used as a component in a variety of pesticides (e. g. algicide, aquatic herbicide, fungicide and molluscicide), which are released in a considerable amount into the environment. The transport of Cu via waste water often leads to enormous burden of aquatic ecosystems, in fact on the one hand to acute and on the other to chronic harmfulness. In 1985 for example limnological investigations of the styrian government showed that the river Mur (Styria, Austria) was intensively burdened by copper. In some sample sites top values of Cu concentrations up to 120 ppb have been measured (Pescheck and Herlicska, 1990). In the meantime the water quality of the Mur is satisfying but the example shows the importance of a reliable biomonitoring system seen from the viewpoint of the applied environmental protection.

## II. 2. Materials and methods

### II. 2. 1. Testorganisms

For our experiments we used laboratory clones of *D. magna* STRAUS (Batch no.: DM300402) hatched in the Laboratory for Environmental Toxicology and Aquatic Ecology (LETAE) at the Ghent University in Belgium and distributed by MICROBIOTESTS Inc. (Belgium). The clones have been delivered in a dormant form (ephippia), thus we were able to activate the test animals on demand, prior to the performance of the toxicity experiments. The ephippia were stored in 1 ml plastic vials, which contained a specific storage medium. The tubes with the ephippia were covered with aluminium foils (dark storage condition) and stored in a refrigerator at 4 °C until use.

II. 2. 1. 1. Hatching of the ephippia

Under optimal conditions the embryonic development of *D. magna* STRAUS takes about three days. Most of the neonates hatch between 72 and 80 hours of incubation. For hatching purposes the content of one vial with ephippia was poured into a microsieve with 100 µm mesh width. To eliminate all traces of the storage medium the ephippia were rinsed thoroughly with tap water. After this procedure the dormant eggs were transferred into hatching petri dishes with 50 ml pre-aerated Standard Freshwater. The Standard Freshwater is an ISO medium and was prepared after the Standard Operating Procedure (SOP) of the Daphtoxkit F$^{TM}$ Magna, which is according to the recommendations of the International Standardization Organization (ISO) for toxicity tests with *D. magna*. In a first step a 2000 ml volumetric flask was filled with 1000 ml distilled water. Then the following concentrated salt solutions (Batch no.: ISOD170201), have been added to the distilled water in the corresponding sequence: 12 ml $NaHCO_3$ → 12 ml $CaCl_2$ → 12 ml $MgSO_4$ → 12 ml KCl. In the last step distilled water has been added again to the salt solution up to the 2000 ml mark of the volumetric flask. Before hatching or doing toxicological experiments the Standard Freshwater was well aerated by means of an aquarium air pump.

The hatching petri dishes have been covered and incubated for 72 hours, at 20 °C under a continuous illumination of 6000 lux in a laboratory incubator (HPS 500, Heraeus Voetsch GmbH, Germany). The actually experiment was conducted with 60 daphnids (10 per group) of the same age - (i. e. neonates ≤ 24 hours old) and body size class. To supply the testorganisms with enough metabolic energy reserves and to avoid bias of the test results due to disturbance of the normal behaviour by starve a 2 hour pre-feeding procedure prior to the toxicity experiments was done with a suspension of Spirulina micro-algae (Batch no.: SP200100).

II. 2. 2. Copper dilutions

In the toxicological experiment the daphnids were exposed to following range of sublethal copper concentrations [in ppb]: $C = 0$, $C_1 = 1$, $C_2 = 5$, $C_3 = 10$, $C_4 = 20$, $C_5 = 30$. The dilutions were produced from a 1000 ppm copper standard solution (Merck). The highest copper concentrations ($C_4$, $C_5$) are based on the range of 48 h $LC_{50}$ values determined on preliminary acute toxicity tests.

II. 2. 3. Measuring of swimming parameters

The design of the ethotoxicological trial is presented in Fig. 3. Ten daphnids of each Cu-treatment were placed into inert polycarbonate test chambers (25 x 25 x 10 mm) under static conditions, i. e. without any water exchange during the experiment.

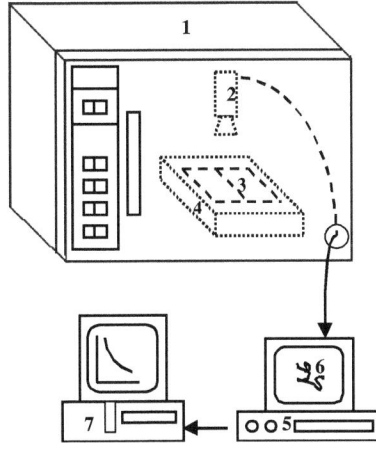

**Fig. 3:** Experimental set-up: 1 laboratory incubator, 2 video camera, 3 test chambers, 4 cold light source, 5 video recorder, 6 monitor showing locomotory activity, 7 Pentium-PC with an integrated frame grabber and analysis software.

The transfer of daphnids was carried out rapidly and carefully by means of a micropipette whilst minimising the stress to the organisms. Above the test cells a video camera (JVC Model no.: GR-SX9E) was installed, which enabled the observation and recording of the behavioural parameters of the test animals. The animals were seen as dark silhouettes on a uniformly illuminated background. The obtained images were digitised, processed and analysed by means of image processing techniques. At first the images were recorded by a conventional video recorder. In a second step the videotapes were digitised using a low-cost frame grabber placed in a Pentium-PC.

The frame grabber consists of an A/D-converter, which enables a digitising of the images in real time (25 frames $s^{-1}$, i. e. $40 \times 10^{-3}$ s interval between two frames).

For further processing of the digitised images the software Image J (Version 1. 29 g) was used. It is a public domain Java image processing program inspired by NIH (National Institute of Health, USA) image for Macintosh. As downloadable application it runs on every PC with a Java 1.1 or later virtual machine. The images were manipulated as series of single images, so called stacks. After thresholding, through the daphnids pixels were set black (grey value = 0) and all other pixels (background) were set white (grey value = 255) and by means of a special Image J plug-in (Multi object tracker), an automated, simultaneous tracking of the daphnids during the whole stack was possible. The temporal resolution between two frames of the stack was set to 40 ms (real-time-analysis). The program records the co-ordinates and automatically reconstructs the trajectories of the daphnids (Fig. 4 b). The data have been stored and exported to a statistical software package for further processing.

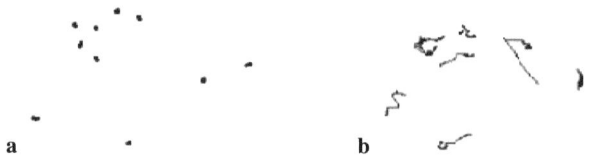

**Fig. 4:** shows the pixels of 10 daphnia's after thresholding (a), (b) shows the reconstructed trajectories during a sequence of 2000 ms.

## II. 2. 4. Statistics

Differences between the observed swimming velocities of the groups of different Cu-treatments were tested with t-tests or by means of a one-way ANOVA followed by Duncan's multiple range post hoc test (P < 0.05). The statistical package SPSS for Windows (Version 9.0, SPSS Inc.) was used for calculating these tests, checking the assumptions of normality (Kolomogorov-Smirnov chi-spuared test) and homoscedasticity (Bartletts chi-spuared test), and calculation of regressions and correlations. A polynomial regression model of the second order ($y = ax^2 \pm bx \pm c$) was applied to determine the $EC_{50}s$ for the behavioural criteria. In comparison with the commonly used linear models a polynomial model resulted in a better fit to the experimental data (Charoy et al., 1995; Charoy and Janssen, 1999).

## II. 3. Results

It could be observed, that the average time of active swimming participation decreased with increasing Cu concentration level, in fact the more the longer the animals were exposed to the Cu-contamination (Fig. 5 a).

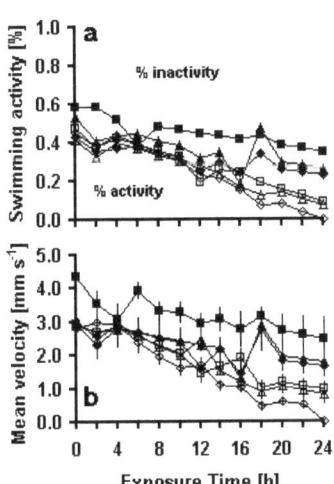

**Fig. 5:** (a) Percentage of the swimming activity during the 24 hour copper exposure. The areas under the graphs show the percentage of active swimming participation. (b) Average swimming velocity (± S. E. of mean = Error bars) of daphnids exposed to copper for 24 hours, plotted against exposure time. ─■─ = 0 ppb (Control), ─▲─ = 1 ppb, ─◆─ = 5 ppb, ─□─ = 10 ppb, ─△─ = 20 ppb, ─◇─ = 30 ppb.

The average swimming velocity decreased with increasing time of exposure (Fig. 5 b) in every group, in fact the more the higher the concentration of Cu contamination.

Tab. 4 shows linear regressions, which describe decreasing of the average velocity in dependence of ongoing exposure time.

**Tab. 4:** Cu-induced decrease of the swimming velocity of *D. magna* STRAUS determined for the trial time of 24 hours. ppb = parts per billion; v = average swimming velocity; ET = Exposure time; $r^2$ = coefficient of determination; P-value = probability value; 95 %-C. I. = confidence interval

| Cu-Treatment (ppb) | Regression | $r^2$ | P-value | 95 %-C. I. |
|---|---|---|---|---|
| 0 | v = -0.0583 ET + 3.8757 | 0.72 | < 0.001 | ± 0.0239 |
| 1 | v = -0.0436 ET + 2.8678 | 0.49 | < 0.01 | ± 0.0293 |
| 5 | v = -0.0466 ET + 2.7852 | 0.49 | < 0.01 | ± 0.0318 |
| 10 | v = -0.0888 ET + 2.9708 | 0.91 | < 0.001 | ± 0.0183 |
| 20 | v = -0.1009 ET + 3.0832 | 0.90 | < 0.001 | ± 0.0225 |
| 30 | v = -0.1283 ET + 3.0860 | 0.96 | < 0.001 | ± 0.0173 |

Within the toxic range used, the 1 ppb and 5 ppb Cu treatment did not significant affect the swimming behaviour of the daphnids (P > 0.05). After 9 hours of Cu stress the average swimming velocity of the daphnids was for the first time significantly reduced from 3.26 mm s$^{-1}$ in the control group to 1.78 mm s$^{-1}$ in the 30 ppb treatment (* P < 0.05). That means a reduction of 54.60 % (Fig. 6). After 13 hours, swimming velocity significantly decreased from 3.13 mm s$^{-1}$ (controls) to 1.65 mm s$^{-1}$ (20 ppb Cu-treatment) (* P < 0.05), i. e. a reduction of 52.72 % (Fig. 6). After 14 hours of copper exposure the swimming velocity was significantly reduced in the 10 ppb Cu-treatment (3.07 mm s$^{-1}$ in the control group versus 1.66 mm s$^{-1}$, i. e. a reduction of 54.07 %, * P < 0.05).

The values of the swimming velocity in Fig. 6 are expressed as percentages relative to the control group.

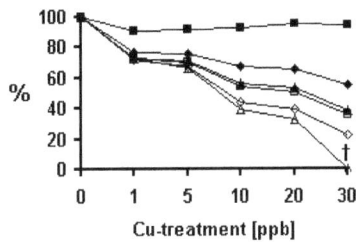

**Fig. 6:** Effects of Cu on the swimming behaviour of *D. magna* Straus expressed as percentages relative to the control. Exposure Time: ■ = 4 h, ♦ = 9 h, ▲ = 13 h, □ = 14 h, ◊ = 20 h, △ = 24 h; †: 100 % immobility after 24 h.

The calculated $EC_{50}$s for the parameter "swimming velocity" were comparable to, or lower than the 24 h $LC_{50}$ values obtained in commonly used acute toxicity tests (Tab. 5).

**Tab. 5:** Calculated $EC_{50}$ values for the behavioural parameter "swimming velocity" at different times of Cu exposure. ET: Exposure time; $v_{dec}$: decreased velocity [%].

| ET [h] | Regression model | $EC_{50}$ [ppb] |
|---|---|---|
| 6 |  | 30.35 |
| 10 |  | 29.09 |
| 14 | $EC_{50} = a\ v_{dec}^2 \pm b\ v_{dec} \pm c$ | 17.03 |
| 20 |  | 10.09 |
| 22 |  | 8.83 |
| 24 |  | 9.77 |

The course of the $EC_{50}$ values in dependence of onwardly Cu exposure time is shown in Fig. 7. It can be described by an exponential regression model ($y = ae^{bx}$; $r^2 = 0.94$).

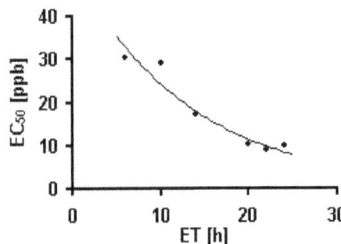

**Fig. 7:** $EC_{50}$ values for the behavioural parameter "swimming velocity" of *D. magna* Straus in dependence of onwardly exposure time (ET). The course can be described by an exponential regression model: $EC_{50} = a\ e^{-bx}$, $r^2 = 0.94$.

In Fig. 8 it is shown that the decrease of swimming velocity correlates with less swimming participation.

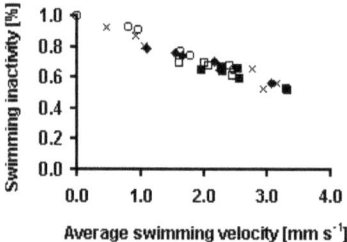

**Fig. 8:** Correlation between the average swimming velocity and swimming inactivity at different times of Cu exposure (ET). ET: ■ 8 h ($r^2$ = 0.79), □ 10 h ($r^2$ = 0.87), ♦ 14 h ($r^2$ = 0.96), × 18 h ($r^2$ = 0.95), ○ 24 h ($r^2$ = 0.97). $r^2$ = coefficient of determination

## II. 4. Discussion

Since it is known that the swimming velocity of daphnids correlates positively with the body size (Dodson and Ramcharan, 1991), Baillieul and Blust (1999) critically pointed out that differences in treatments will reflect differences in body size rather than real differences in swimming velocity. In their studies they found out, that sublethal cadmium stress decreases growth of *D. magna* STRAUS. On the one hand we considered this point by using standardised testorganisms of the same age- and body size class. On the other hand our experiment was performed in a time span (24 h), in which bias of the results on swimming velocity due to effects of the body size (growth) couldn't occur.

However, according to Wolf et al. (1998) we could observe that the daphnids needed an adaptation period of ≥ 2 h before showing a constant swimming behaviour. Thereafter analysing the swimming parameters was useful and was performed consequently in our experiments. Our results show that ethotoxicological biomonitoring systems, using the parameters "swimming velocity" and "- participation" as biological endpoints are to evaluate as very sensitive for detecting sublethal Cu stress. This conclusion is in accordance with results of other investigators, who measured behavioural parameters after submission of daphnids to sublethal xenobiotica stress (Knie, 1978; Wolf et al., 1998; Baillieul and Blust, 1999).

Sublethal effects of Cu on behavioural activity have already been reported for fish at $\geq 10~\mu g~l^{-1}$ (Lett et al., 1976; Waiwood and Beamish, 1978; Beitinger and McCauley, 1990), bivalves at 5 µg $l^{-1}$ (Salanki, 1992; Mersch et al., 1993), amphipodes at $\leq 50~\mu g~l^{-1}$ (Gerhardt, 1995) and mysids (Garnacho et al., 2001). A decreased locomotion and an increased ventilation of *Gammarus pulex* L. (Crustacea) exposed to sublethal Cu concentrations were shown by Gerhardt (1995). The feeding activity of the snail *Campeloma decisum* on clam meat ceased at copper concentration of 0.015 and 0.028 mg $l^{-1}$ and the operculum closed at higher concentrations (Arthur and Leonard, 1970). The freshwater snail *Thiara tuberculata* reduced locomotion when exposed to 5 mg $l^{-1}$ Cu by withdrawing inside its shell. Exposure to a Cu concentration of 1 mg $l^{-1}$ resulted in normal movement for the first 10 days and a gradual reduction up to 20 days. Furthermore the oxygen consumption was severely reduced (Mule, Lomte, 1994).

Since locomotion needs the coordination of the nervous system, it is supposed, that Cu has integrative effects on the nervous system and muscular systems. Neurophysiological effects of Cu were shown for different aquatic invertebrates (Salanki, 1992; Ferrando and Andreu, 1993; Gerhardt, 1995). For example, Cu affects the permeability of neuronal membranes in bivalves (Salanki, 1992) and depresses acetylcholine activity in the synapses of *Chironomus decorus* (Kosalwat and Knight, 1987). S.-Rozsa and Salanki (1992) observed that Cu modified the effect of neurotransmittors and ionic current in the simple nerve systems of the pond snail *Lymnaea stagnalis*. Findings of Baillieul and Blust (1999) showed, that the beat frequency of the second antennae, which are responsible for swimming activity of *D. magna* decreased with increasing cadmium concentration due to neurological failure. Since the complex pattern of the locomotory behaviour can be considered as an integration of physiological, sensorial, nervous and muscular systems (Charoy et al., 1995), findings of several papers (Knops et al., 2001; Heath, 1995) suggest that exposure of aquatic organisms (invertebrates and vertebrates) to sublethal copper concentrations causes increased maintenance costs resulting in higher metabolic rates in certain non-muscular tissues while spontaneous muscular activity becomes depressed. Thus, a chronic copper stress induces a loss in the production of metabolic energy for muscle activity and as a consequence for locomotion.

The results of the present study show that a rising copper stress induces a measurably decrease of the swimming velocity, which seems to be a very sensitive biological endpoint. According to Wolf et al. (1998) the observed behavioural patterns, which our daphnids show under sublethal copper stress are to interpret as typical protection reaction, characterised in a first line by a decreased locomotory activity due to the loss of coordination.

As far as it is concerning the question after what time span a reliable evaluation of copper burden is possible, our results show that we can obtain significant values in a time range from 9 to 15 hours of Cu exposure (depending on the concentration level). Compared to the results of acute toxicity tests (24 and 48 h $LC_{50}$), which measure the parameter "mortality" as biological endpoint, the image analysing biomonitoring system can be classified as more sensitive and superior as the results can be obtained very fast and the Cu induced stress can be detected at sublethal concentration levels. Our calculated $EC_{50}$ values (Tab. 5) for the behavioural parameter "swimming velocity" are comparable to or more sensitive than 24 (48) h $LC_{50}$ values of commonly applied acute toxicity tests, respectively.

## II. 5. Conclusions

This study proved behavioural responses of the freshwater cladoceran *D. magna* STRAUS to be sensitive bioindicators to copper stress. Changes in the behavioural parameters "swimming velocity" and "- participation" can be used as early stress responses for chronic Cu contamination as part of ecological risk assessment.

# Chapter III – *Daphnia pulex* LEYDIG

Toxic effects of the heavy metal Cu on the locomotory behaviour of *Daphnia pulex* Leydig (Crustacea: Cladocera) - Analysis by means of video trajectometry

with Jördis Kahapka

Abstract

In this study the sublethal toxicity of the heavy metals copper to *Daphnia pulex* Leydig has been investigated. Sublethal effects were evaluated using changes in the locomotory behaviour (i. e. swimming velocity and – activity) as indicators. The locomotory activity were analysed by means of real time image analysis, using a video camera and a Pentium-PC equipped with a standard low cost frame grabber. For a sequence of 50 images per treatment, where 10 daphnids were moving simultaneously, the trajectories have been automated reconstructed in binary image sequences. The locomotory activity of the test organisms were analysed under normal conditions (without heavy metal stress) and after application of sublethal Cu stress. Test animals were stressed by Cu of following concentrations: $C_1$ = 5 ppb, $C_2$ = 10 ppb. The daphnids were exposed to the heavy metal concentrations for 24 hours under static conditions. Already after 9 hours of Cu-exposure test animals showed a very high significant (*** $P < 0.001$; ) decrease of the average swimming velocity at $C_1$ and a high significant (** $P < 0.01$) decrease at $C_2$.

*Key words:* Sublethal toxicity, Ethotoxicology, Copper, Daphnids

## III. 1. Introduction

In scientific biomonitoring it is common to use different species of anomopods like *Daphnia magna* Straus and *Daphnia pulex* Leydig as testorganisms. In most cases the acute toxicity of chemical compounds are determined using the parameter mortality as biological endpoint. To determine sublethal effects of chemical substances it is common to use behavioural or physiological parameters as biological endpoints, since the behaviour of an organism is defined as the end point of a sequence of different neurophysiological processes (Lagadic et al., 1994; Gerhardt, 1995; Untersteiner et al., 2003). Especially the complex pattern of locomotory behaviour can be considered as an integration of physiological, sensorial, nervous and muscular systems (Charoy et al, 1995; Untersteiner et al., 2003). Thus, besides physical-chemical water analysis, analysing of the locomotory stress behaviour is an important complementary tool for evaluating the toxic potential of xenobiotica in classical biomonitoring.

## III. 2. Material and methods

III. 2. 1. Test organisms

For the ethotoxicological trials we used laboratory clones of *D. pulex* Leydig (Batch no.: DP) hatched in the Laboratory for Environmental Toxicology and Aquatic Ecology (LETAE) at the Ghent University in Belgium and distributed by MICROBIOTESTS Inc. (Belgium). The clones have been delivered in a dormant form (ephippia) and have been stored in 1 ml plastic vials, which contained a specific storage medium. The tubes with the ephippia were covered with aluminium foils and stored in a refrigerator at 4 °C until use.

III. 2. 1. 1. Hatching of the daphnids

Hatching of *D. pulex* ephippia was done after the same procedure as described under II. 2. 1. 1. for *D. magna*.

### III. 2. 2. Heavy metal dilutions

In the ethotoxicological experiment the daphnids were stressed by following range of sublethal copper concentrations [in ppb]: $C = 0$, $C_1 = 5$, $C_2 = 10$, $C_3 = 20$. The dilutions were produced from a 1000 ppm copper standard solution (Merck).

### III. 2. 3. Measuring of the locomotory activity

The design of the ethotoxicological experiment is presented in Fig. 3. The method is the same as for *D. magna* and is already described under II. 2. 3.

## III. 3. Results

It could be observed, that the animals of the highest Cu treatment died immediately after the transfer in the Cu solution, therefore only the treatment groups C, $C_1$ and $C_2$ have been taken for further analysis.

In general it could be observed that the average swimming velocity of *D. pulex* Leydig decreased with increasing time of exposure (Fig. 9).

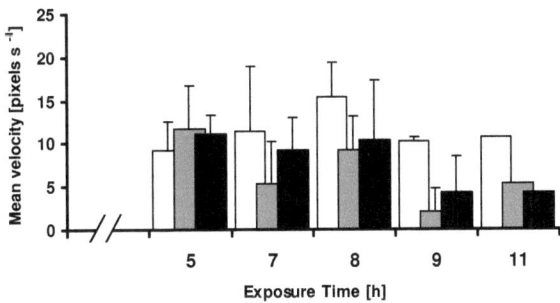

**Fig. 9:** Average swimming velocity (± S. D. = Error bars) of *D. pulex* exposed to copper for ≤ 11 hours, plotted against exposure time. ☐ 0 ppb (Controls), ▨ 5 ppb; ■ 10 ppb.

Tab. 6. shows linear regressions, which describe decreasing of the average velocity in dependence of ongoing exposure time.

**Tab. 6:** Cu-induced decrease of the swimming velocity of *D. pulex* Leydig

| Cu-treatment [ppb] | Regression | $r^2$ | P-value |
|---|---|---|---|
| 0 | v = -0.1246 ET + 2.5558 | 0.36 | < 0.05 |
| 5 | v = -0.1760 ET + 2.3163 | 0.73 | <0.001 |
| 10 | v = -0.1783 ET + 2.3942 | 0.84 | <0.001 |

Legend: ppb = parts per billion; v = average swimming velocity; ET = exposure time; $r^2$ = coefficient of determination; P-value = probability value.

After 9 hours of Cu stress the average swimming velocity of the daphnids was for the first time very high significantly reduced from 1.22 mm s$^{-1}$ in the control group to 0.26 mm s$^{-1}$ in the 5 ppb treatment (*** P < 0.001). That means a reduction of 79.11 % (Fig. 10 a). At the same time of Cu exposure a high significant decrease from 1.22 mm s$^{-1}$ in the control group versus 0.51 mm s$^{-1}$ in the 10 ppb group (** P < 0.01) could be observed. That means a reduction of 58.08 % (Fig. 10 a). After 11 hours of Cu-exposure a significant decrease could be observed in the 5 ppb group (1.28 mm s$^{-1}$ in the control group versus 0.64 mm s$^{-1}$, i. e. a reduction of 50.02 %, * P < 0.05) and a high significant decrease could be measured in the 10 ppb group (1.28 mm s$^{-1}$ in the controls versus 0.52 mm s$^{-1}$, i. e. a reduction of 59.26 %, ** P < 0.01), respectively (Fig. 10 b).

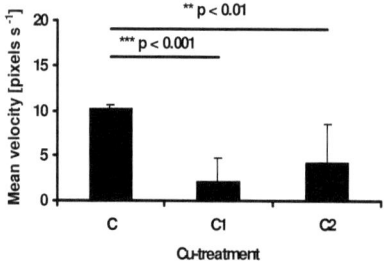
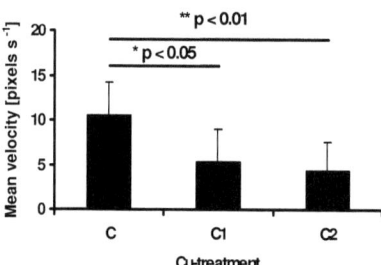

**Fig. 10:** Decreased swimming velocity of *D. pulex* Leydig due to Cu stress. (a) 9 h Cu-exposure, (b) 11 h Cu-exposure. C = Control, $C_1$ = 5 ppb Cu-treatment, $C_2$ = 10 ppb Cu-treatment.

The calculated $EC_{50}s$ for the parameter "swimming velocity" were comparable to or lower than the 24 h $LC_{50}$ values, respectively obtained in commonly used acute toxicity tests (Tab. 7).

**Tab. 7:** Calculated $EC_{50}$ values for the behavioural parameter "swimming velocity" at different times of Cu exposure

| ET [h] | Regression model | $EC_{50}$ [ppb] |
|---|---|---|
| 4  |   | 7.76 |
| 6  |   | 6,95 |
| 7  | $EC_{50} = av^2_{dec} \pm bv_{dec} \pm c$ | 6,73 |
| 9  |   | 5,37 |
| 10 |   | 5,33 |
| 11 |   | 5,11 |

Legend: ET: = exposure time: $v_{dec}$ = decreased velocity (%)

The course of the $EC_{50}$ values in dependence of onward Cu exposure time is shown in Fig. 11.

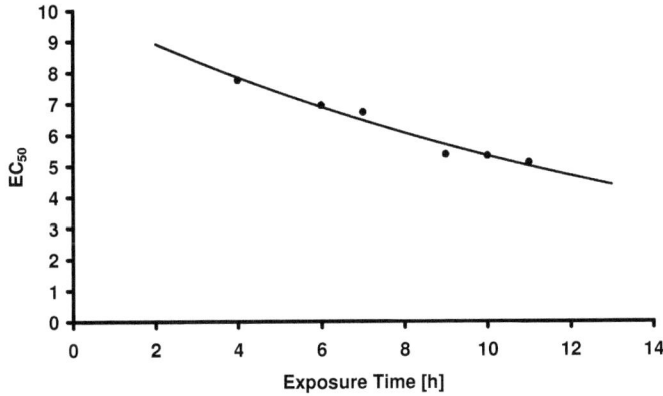

**Fig. 11:** $EC_{50}$ values for the behavioural parameter "swimming velocity" of *D. pulex* LEYDIG in dependence of onwardly exposure time (ET). The course can be described by an exponential regression model: $EC_{50} = a(e^{-bx})$, $r^2 = 0.96$.

## III. 4. Discussion

Like in the trial with *D. magna* only test organisms of the same age- and body size class have been used, to avoid possible effects of differents body sizes on the swimming velocity. Influences of the body size on the swimming velocity have been discussed by Dodson and Ramcharan, 1991 and by Baillieul and Blust (1999).

The results of this trial show that *Daphnia pulex* is a high sensitive organism and very suitable for monitoring heavy metal stress. The used behavioural parameter swimming velocity and – activity are very sensitive for detecting sublethal copper stress. The obtained $EC_{50}$-values (Tab. 7) are to interpret as more sensitive than the $LC_{50}$ values. The $LC_{50}$- values of our test animals ranges from 15 to 20 ppb, while the obtained $EC_{50}$s for the parameter ranges from 5 to 10 ppb, depending on the exposure time.

Physiological effects, which lead to these behavioural ataxias, have been intensively discussed in the trial of *Daphnia magna* (II. 4.). Since the physiological, biochemical and behavioural mechanisms of *D. pulex* are similar to those of *D. magna* the interpretation can be transferred to this trial.

However, differences in the Cu sensitivity and behavioural responses have been selected and intensively discussed under V. 2.

## III. 5. Conclusions

This study proved behavioural responses of the freshwater cladoceran *D. pulex* LEYDIG to be sensitive bioindicators to copper stress. Changes in the behavioural parameters "swimming velocity" and "- participation" can be used as early stress responses for chronic Cu contamination as part of ecological risk assessment.

# Chapter IV – *Hippolyte inermis* LEACH

**Monitoring behavioural responses to the heavy metal Cd in the marine species *Hippolyte inermis* LEACH (Crustacea: Decapoda) by means of video imaging**

with Gerwin Gretschel, Tom Puchner, Sonja Napetschnig

## Abstract

In this study the sublethal toxicity of the heavy metal cadmium to *Hippolyte inermis* LEACH has been investigated. Sublethal effects were evaluated using changes in the locomotory behaviour (i. e. moving velocity and – distance) as indicators. The locomotory activity was analysed by means of real time image analysis, using a video camera and a Pentium-PC equipped with a standard low cost frame grabber. For a sequence of 3000 images per treatment, where 10 shrimps were moving simultaneously, the trajectories have been reconstructed in binary image sequences. The locomotory activity of the test organisms was analysed under normal conditions (without heavy metal stress) and after application of sublethal Cd stress. Test animals were stressed by Cd of following concentrations: $C_1 = 1$ ppm, $C_2 = 2$ ppm, $C_3 = 3.5$ ppm. The shrimps were exposed to the heavy metal concentrations for 12 hours under static conditions. Already at the start time (0 h) of Cd-exposure test animals showed a high significant (** $P < 0.01$; ) decrease of the average swimming velocity at $C_3$. After 1 h of Cd exposure, median moving velocity significantly decreased at $C_2$ (* $P \leq 0.05$). After 3 hours of Cd exposure, median moving velocity was for the first time reduced with high significance in the 1 ppm Cd-treatment (** $P \leq 0.01$).

*Key words:* Sublethal toxicity, Ethotoxicology, Cadmium, Hippolytidae

## IV. 1. Introduction

The Adriatic Sea is a complex and sensitive ecosystem in which the different organisms can be exposed to various anthropogenous associations of chemical compounds. The northern Adriatic Sea is strongly influenced by freshwater discharges from the Po and adjacent rivers (Tankere and Statham, 1996). Although the Adriatic Sea is not as heavily contaminated by Cd-pollution as other marine ecosystems like the North Atlantic Ocean, North Pacific Ocean, Indian Ocean, Baltic or Black Sea (Sekulić and Vertačnik, 1997) investigations from Tankere and Statham (1996) showed that the Po-influenced region of the Adriatic (northern part) shows higher Cd concentrations than the central and southern region. The major source for the heavy metal Cadmium (Cd) is likely to be the river Po. It is known that Cd can be mobilized from sediment under reducing environmental conditions and consequently diffuses from pore waters to the water column (Zago et al., 2000).

The toxic potential of Cd is enormous, since continuous exposure of marine organisms to an already low concentration of Cd may result in bioaccumulation, and subsequent transfer to man through the food web (Kljaković Gašpic et al., 2002). As a result of the global concern over impact of xenobiotica on aquatic ecosystems and human health, several scientific biomonitoring systems have been developed using different invertebrates and vertebrates as indicator organisms (e. g. Fish-Rheotaxis-test, Koblenz behavioural fish test, Dreissena-Monitor and the Dynamic Daphniatest) (Knie, 1978; Bocherding, 1992; Schmitz et al., 1994; Gerhardt, 1999; Untersteiner et al. 2003). In Europe and the United States of America numerous bioanalytical techniques for the control of water quality have been developed during the past decades and applied at the sub- and multicellular levels of biological organization (Tab. 1).

In most cases of scientific biomonitoring the acute toxicity of chemical compounds is determined using mortality rates as parameter. To determine sublethal effects of chemical substances it is common to use behavioural or physiological parameters as biological endpoints, since the behaviour of an organism is defined as the endpoint of a sequence of different neurophysiological processes (Lagadic et al., 1994; Gerhardt, 1995; Untersteiner et al., 2003). Especially the complex pattern of locomotory behaviour can be considered as an integration of physiological, sensorial, nervous and muscular systems (Charoy et al, 1995; Untersteiner et al., 2003).

Although a lot of publications exist which are dealing with Cd-toxicity to different aquatic invertebrate and vertebrate species (Viarengo et al., 1997; Fichet et al; 1998; Rasmussen and Andersen, 2000; Rainbow et al., 2000; Adami et al., 2002; Kljaković Gašpic et al., 2002; Filipović and Raspor, 2003) no one has studied the Cd-sensitivity of *Hippolyte inermis* Leach (Crustacea: Decapoda). This organism meets various requirements and therefore appeared to be potentially suitable for biomonitoring. It is very abundant in an easily accessible and ecologically extremely important habitat i. e. seagrass meadows of *Posidonia oceanica* and *Cymodocea nodosa* (Gambi et al., 1992). It can be sampled easily and in high numbers by divers and it can be reared in tanks in good condition and a sufficient time period to carry out experimental studies (Zupo, 2000).

The aim of this work was therefore to study the acute and sublethal toxicity of the trace metal Cd to the marine crustacean species *H. inermis* and to test it's suitability as test organism in the context of classical biomonitoring.

## IV. 2. Material and methods

IV. 2. 1. Study sites and test organisms

The test organisms were collected from a seagrass habitat (*Cymodocea nodosa* (UCRIA)) in the bay of Valsaline, Pula, Croatia (44° 51′ 044″ N / 13° 50′ 080″ E) (Fig. 12).

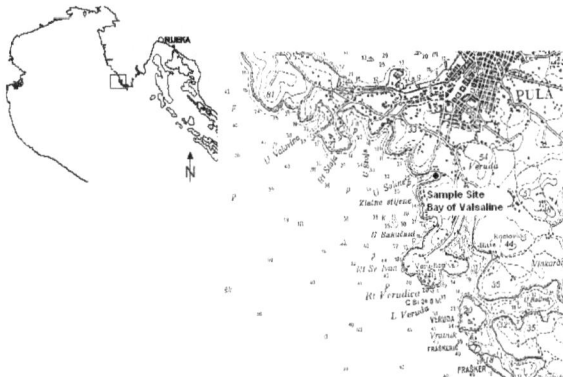

**Fig. 12:** Location of the sampling sites in the bay of Valsaline – Pula (44° 51′ 044″ N / 13° 50′ 080″ E)

For every toxicological trial specimens were sampled by two divers from a depth of 4 – 6 m by pulling a simple hand dredge (Frame: 15 x 20 cm, net length: 105 cm, mesh size: 1 mm) through the seagrass meadow (10 – 15 cm above ground). Sampling by divers proved to be highly efficient (~ 15 - 30 individuals per minute). Impact on the seagrass habitat was minimal compared to conventional dredging by boat. The test animals were immediately transferred into tanks and sorted by body size. For the toxicological trial test animals of the same body size class (~ 1.4 cm total length from the tip of the rostrum, to the posterior medial Notch) have been used. The acute toxicity tests (range finding and definitive test) have been conducted with 100 shrimps (20 per group), the ethotoxicological trials have been conducted with 40 animals (10 per group).

To assess the natural pollution with cadmium a water sample was taken for chemical analysing in the laboratory (Tab. 8).

**Tab. 8:** Physico-chemical parameters of the seawater sample taken in the bay of Valsaline.

| Parameter | Values |
|---|---|
| Cd µg/l | $0.46 \pm 0.01$ |
| Ca mg/l | $469 \pm 4$ |
| Mg mg/l | $1340 \pm 20$ |
| °dH | 375 |
| pH | 7.4 |
| Salinity | $36 \text{ g kg}^{-1}$ |

IV. 2. 2. Cadmium dilutions

A 36 ‰ standard seawater solution (Aqua marina) was prepared for the toxicological tests. For a time period of one hour before doing toxicological experiments the standard seawater solution was well aerated by means of an aquarium air pump. In the ethotoxicological experiment the test organisms were stressed by following range of sublethal cadmium concentrations [in mg $l^{-1}$]: C = 0, $C_1$ = 1, $C_2$ = 2, $C_3$ = 3.5. The dilutions were produced from a 1000 ppm cadmium standard solution (Merck). The highest Cd-concentration ($C_3$) is based on the range of 12-h-$LC_{50}$ values determined on preliminary acute toxicity tests.

### IV. 2. 3. Measuring of locomotory activity

The design of the ethotoxicological experiment is presented in Fig. 13. The equipment was placed into a climatic room to keep the temperature constant to 27 ± 0.5 °C during the trials.

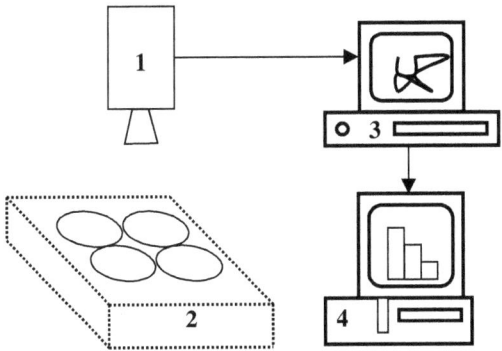

**Fig. 13:** Experimental set-up: 1, video camera; 2, pad with petri dishes; 3, video recorder with monitor showing locomotory behaviour; 4, Pentium-PC with an integrated frame grabber and analysis software.

Ten shrimps of each Cd-treatment were placed into inert petri dishes (Ø 85 mm) under static conditions, i. e. without any water exchange during the trial. The transfer of the test organisms was carried out carefully by means of a micropipette whilst minimising the stress to the organisms. Above the petri dishes a video camera (JVC model no.: GR-SX9E) was installed, which enabled the observation and recording of the behavioural parameters of the test animals. The animals were seen as dark silhouettes on a uniformly illuminated background (Fig. 14). The obtained images were digitised, processed and analysed by means of image processing techniques. This process has already been described in II. 2. 3.

# Chapter IV – *Hippolyte inermis* LEACH

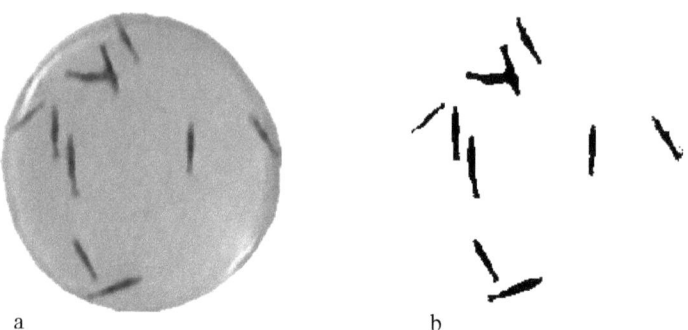

a                               b

Fig. 14: (a) Shows the silhouettes of 10 seagrass shrimps, (b) shows the pixels of the shrimps after thresholding.

IV. 2. 4. Statistics

The 12h-$LC_{50}$-values have been calculated by means of probit analysis. Since the collected data show no normal distribution, differences between the observed locomotory activities (moving velocity, distance moved) were tested with H-Test after Kruskal and Wallis ($P < 0.05$). In case of significant results of the H-Test, differences between 2 groups were tested by means of an U-Test after Mann and Whitney in each case. The validity of an ethotoxicological trial was tested by means of a survival analysis (Fig. 15). As criteria we defined a survivability of 90 percent of the test animals in the controls during the trial. The statistical package SPSS for Windows (Version 9.0; SPSS Inc.) was used for calculating these tests.

IV. 3. Results

The chemical analysis of the water sample showed that the research site (bay of Valsaline) is less polluted with Cadmium (Tab. 8) thus we could assume, that our research organisms were not adapted to Cd. As a result of the acute toxicity tests (range finding and definitive test) a 12 h-$LC_{50}$-value from 3.45 mg $l^{-1}$ Cd has been calculated. Tab. 9 shows an overview about the toxicological data calculated for *H. inermis* LEACH.

## Chapter IV – *Hippolyte inermis* LEACH

**Tab. 9:** Calculated 12 h-LC-values for the biological endpoint "mortality"

|  | LC-Value [ppm] | 95% confidence limits [ppm] | |
|---|---|---|---|
|  |  | lower | upper |
| 12 h-$LC_{10}$ | 1.26 | 0.62352 | 2.09784 |
| 12 h-$LC_{50}$ | 3.45 | 2.67498 | 4.61758 |
| 12 h-$LC_{90}$ | 5.64 | 4.51534 | 8.36748 |

The ethotoxicological trial with *H. inermis* Leach was defined as valid, when the animals in the controlgroup showed a survivability of ≥ 90 %. To determine this survivability a survival analysis was done. The results are shown in Fig. 15.

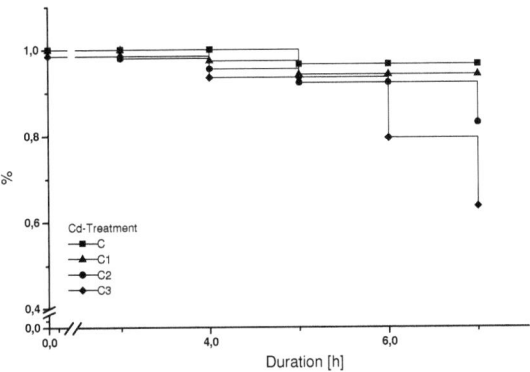

**Fig. 15:** Survival analysis for assessing the validity for the ethotoxicological trial. A trial was valid, when the survivability in the controls was at least 90 %. C = 0 ppm Cd, $C_1$ = 1.0 ppm Cd, $C_2$ = 2.0 ppm Cd, $C_3$ = 3.5 ppm Cd. Ordinate (y-axis) = survivability [%], Abszissa (x-axis) = Exposure Time [h]

Already at the start time (0 h) of the ethotoxicological trial the swimming activity of the group of 3.5 ppm Cd-Treatment showed a high significant difference to the controls. The median moving velocity was reduced with high significance from 0.63 mm $s^{-1}$ (controls) to 0.00 mm $s^{-1}$ (3.5 ppm Cd-Treatment) (** $P \leq 0.01$). The median distance moved was reduced with high significance from 75.74 mm (controls) to 0.00 mm in the 3.5 ppm Cd-Treatment (** $P \leq 0.01$). After 1 h of Cd exposure, median moving velocity significantly decreased from 0.34 mm $s^{-1}$ (controls) to 0.00 mm $s^{-1}$ in the 2 ppm Cd-treatment (* $P \leq 0.05$). After 3 hours of Cd exposure, median moving velocity was for the first time reduced with high significance in the 1 ppm Cd-treatment (0.25 mm $s^{-1}$ in the control group versus 0.06 mm $s^{-1}$, ** $P \leq 0.01$).

Fig. 16 shows an overview about the calculated descriptive data of the ethotoxicological trial.

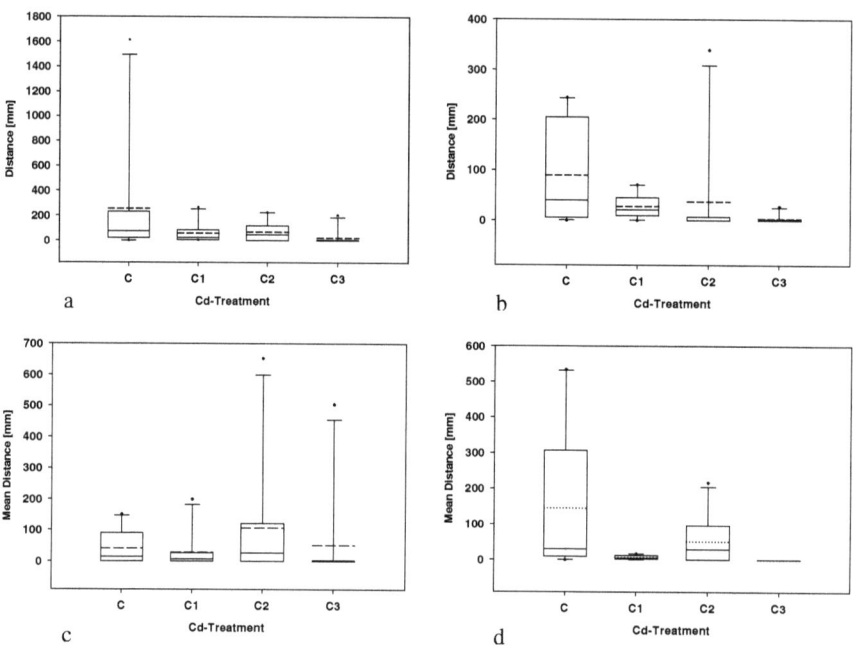

**Fig. 16:** Effects of Cd on locomotory activity (Distance moved) of *H. inermis* LEACH at different times of Cd-Exposure. (a) = 0 h, (b) = 1 h, (c) = 2 h, (d) = 3 h. Data are depicted as Median (solid line), Mean (dashed line), $25^{th}$ and $75^{th}$ percentile (boundary of boxes), $10^{th}$ and $90^{th}$ percentile (whiskers).

## IV. 4. Discussion

In general *H. inermis* reacts very sensitive to artificial environmental conditions. After a time period of 12 hours it is not possible to make reliable ethotoxicological trials under static conditions, because the mortality of the controls increases very rapidly. The presumed cause is the relatively high water temperature of 27°C compared to 18°C used by Zupo (2000). However, for a time period ≤ 12 hours the ethotoxicological results are reliable (Fig. 15).

It could be observed that our test organisms reacted very sensitive to cadmium stress. Already during a time period of 3 hours we could measure Cd effects (Fig. 16). The survivability decreases with increasing Cd-concentration (Fig. 15).

From the observed data we could derive following processes of the stress response of *H. inermis* Leach (Fig. 17): an adaptation reaction, a protection reaction and escape reaction of a first and second order.

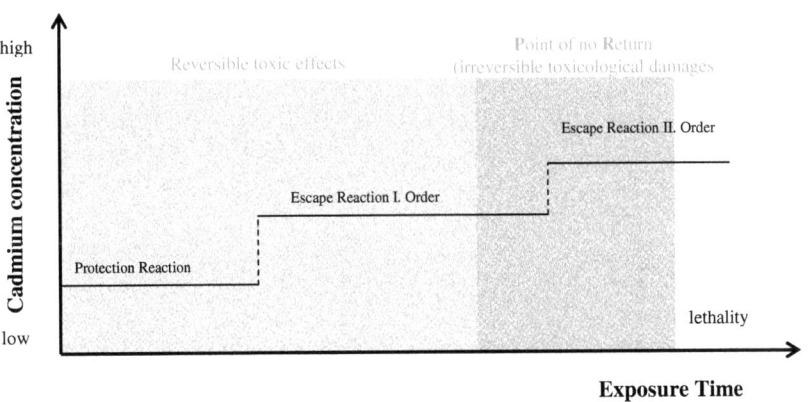

**Fig. 17:** Stress Response Model from *H. inermis* Leach due to Cd-stress. With increasing toxicant concentration, the organism switches from one behaviour to the next behaviour according to their respective range of regulation.

Chapter IV – *Hippolyte inermis* LEACH

During an adaptation reaction a part of the metabolism energy will be used to restore this imbalance.

The protection reaction is characterised by a decreased motility. During this phase a spontaneous muscular activity becomes depressed due to higher maintenance costs resulting in higher metabolic rates in certain non-muscular tissues (Knops et al., 2001; Heath 1995).

An escape reaction of the first order is characterised by an increased ventilation and motility, whereby the escape reaction of the second order is characterised by an very high increased ventilation activity and motility due to powerful beats of the pleon. This energy mobilisation during an escape reaction of the second order is the last chance for the animals to escape unfavourable environmental conditions, otherwise they will die (irreversible toxicological effects). A protection reaction was shown by the test animals exposed to lower Cd-concentrations ($C_1$). Test animals of the $C_2$-group also showed a protection reaction in the first time of Cd-Exposure (0, 1 h). After 2 hours of Cd-Exposure the test animals of the $C_2$-group showed an typical escape reaction of the first order (Fig. 18). In this phase they mobilised metabolism energy for increased muscle activity. Similar stress response mechanisms could be described by other crustaceans, like daphnids (Wolf et al., 1998; Untersteiner et al., 2003).

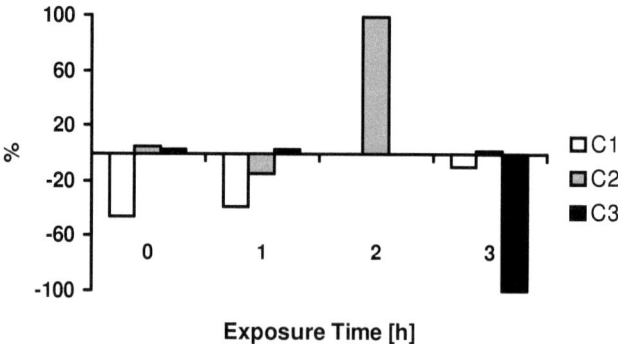

**Fig. 18:** Effects of Cd on the motility of *H. inermis* LEACH expressed as percentages relative to the control.

The locomotory behaviour plays an important role in the evaluation of toxic compounds released into an ecosystem, since the complex pattern of locomotory behaviour can be considered as an integration of physiological, sensorial, nervous and muscular systems (Charoy et al., 1995).

It is known that heavy metals influence a broad spectrum of physiological processes in organisms. Especially effects of heavy metals on the nervous system are very important since the nervous system regulates and coordinates the locomotory behaviour. Findings of Baillieul and Blust (1999) showed, that the beat frequency of the second antennae, which are responsible for swimming activity of *D. magna* STRAUS decreased with increasing cadmium concentration due to neurological failure. A correlation between heavy metal concentration and toxic neurological effects in mussels has been shown by Salanki (1992).

Other physiological effects of the non-essential metal Cd have been discussed by several authors. Effects of Cd on changes in genetic variability and allele frequency on the population level of the gastropod *Littorina brevicula* have been investigated by Kim et al., (2003). Viarengo et al., (1997) used the encyme Metallothionein of Mediterranean and Antarctic molluscs *Mytilus galloprovincialis* and *Adamussium colbecki* as indicator for heavy metal stress. Metallothioneins are metalbinding proteins whose neosynthesis represents a specific response of different organisms to heavy metal stress. An increasing heavy metal concentration in the cells stimulates the synthesis of apothioneins that can bind metal cations in a non-toxic form, thus reducing their deleterious effects (Viarengo, 1997). It can be assumed that metallothioneins play an important role during adaptation- and protection reactions of organisms.

Our results showed that *H. inermis* reacts very sensitive to Cd-pollution. When sampled at seasons with lower water temperature and reared at temperatures around 18°C to increase longevity under artificial conditions, *H. inermis* could prove useful as standardised test organism for biomonitoring.

# Chapter V – General Discussion

## V. 1. Video analysis

Video techniques are common methods in the zoological ethology and ethotoxicology (Wratten, 1994). The experimental design (Fig. 3, 13) used for the ethotoxicological trials are suitable tools for analysing behavioural parameters (biological endpoints) of our testorganisms.

### V. 1. 1. Analysis software

The used image analysis software Image J is most suitable for analysing behavioural parameters of animals. In Image J, Videos are manipulated as so called stacks. Recorded videos have to be thresholded, i. e. the background must be substracted from the pixels of interest. This procedure demands a carefully videorecording to avoid the problem of a non uniformly illuminated background.

Some points highlight the great advantage of Image J as scientific image analysis software:

- It is a free downloadable, public domain program based on the programming language Java.

- There exists already a lot of valuable analysis tools (plug ins) for free download in the internet. (In this work the automated analysis was achieved by means of the plug in "Multi Object Tracker").

- Every scientist can adapt the program for the special analysis purpose of the research work.

- Videos manipulated as stacks requires less storage space on the harddisk of the computer.

## V. 2. Copper effects on daphnids

In general it could be observed that both *Daphnia* species (*D. magna* and *D. pulex*) are very good indicators for Cu- stress. However, when comparing the sensitivity of both organisms it can be observed, that the used neonates of *D. pulex* are more sensitive to Cu than the neonates of *D. magna*. The 24-h $LC_{50}$-value for the used *Daphnia* clones amount to 30 µg l$^{-1}$ Cu for *D. magna* versus 15 – 20 µg l$^{-1}$ Cu for *D. pulex*. Comparing the $EC_{50}$s between this two species, it can be observed, that our clones of *D. pulex* react about 4 times higher to sublethal than our used clones of *D. magna* (Tab. 5; Tab. 7). In order to compare the sensitivity of both *Daphnia* species used in our trials we used the same copper species and the same method for determining the sublethal effects of copper.

However, it is known that a direct comparison of toxicological data published in different scientific paper is difficult and depends on the species used and their genetically constitution. Therefore it is no wonder, that the scientific literature differs in their interpretation of (E)$LC_{50}$-values. Seen from this point of view the paradigm of the most sensitive species as the best environmental indicator is controversial, because species sensitivity is a function of a number of interrelated factors (e. g. environmental conditions, physiological constitution, competitive interactions). Cairns (1986) therefore stated, that predictive capacity rather than sensitivity may be the most important issue to consider when assessing the ultimate utility of any bioassay method.

A composition of acute toxicity data ($LC_{50}$-values) for *D. magna* and *D. pulex* after different authors is shown in Tab. 10.

**Tab. 10:** Acute copper toxicity data on daphnids.

| Species | Lifestage | Test Type | $LC_{50}$-value [mg Cu l$^{-1}$] | Reference |
|---|---|---|---|---|
| *D. magna* | n.d. | FT | 48-h $LC_{50}$ = 0.020 | Bishop, Perry, 1981 |
| *D. magna* | < 24 h | SS (1/2d); CuCl$_2$ | 96-h $LC_{50}$ = 0.130 | Blaylock et al., 1985 |
| *D. magna* | < 24 h | S; CuO | 48-h $LC_{50}$ = 0.026 | Lewis, 1983 |
| *D. pulex* | 24 ± 10 h | CuNO$_3$ | 48-h $LC_{50}$ = 0.003 | Jop et al., 1993 |
| *D. pulex* | < 24 h | S | 48-h $LC_{50}$ = 0.037 | Dobbs et al., 1994 |
|  |  |  | 72-h $LC_{50}$ = 0.026 |  |
| *D. pulex* | < 24 h | S | 72-h $LC_{50}$ = 0.028 | Winner, 1985 |
|  |  |  | 72-h $LC_{50}$ = 0.028 |  |

Legend: Test Type: S = Static, SS = Semistatic, FT = Flow Through
Lifestage: n.d. not described; < 24 h = neonates < 24 h old.

Chapter V – General discussion

It could be observed, that *D. magna* and *D. pulex* differs in their stress response mechanisms as result of copper stress. *D. magna* shows a negatively correlation regarding the swimming activity and –velocity to increasing copper concentration (Fig. 5).

This behaviour is mainly to interpret as typical protection reaction. In the Cu-concentration used in the experiment (1 to 30 ppb) no exemplary escape reaction is shown. After 18 hour of continous Cu-exposure, velocity of the animals in the $C_1$-group (1 ppb) and in the $C_2$-group (5 ppb) approach to the velocity of the controls (Fig. 5). This behaviour could carefully be interpreted as escape reaction. In the groups of higher Cu-concentrations it seems that metabolism energy is used for inner adaptional mechanisms and not for increased locomotory activity.

*D. pulex* shows an increased swimming activity and –velocity in the $C_2$ (10 ppb)-group (velocity = 0.51 mm $s^{-1}$) in comparison to the $C_1$ (5ppb)-group (velocity is 0.26 mm $s^{-1}$). Although both velocities ($C_1$ and $C_2$) are significantly lower than the average velocity in the control group (1.22 mm $s^{-1}$) the higher velocity in the $C_2$-group is to interpret as typical escape reaction. In this phase of stress response individuals of *D. pulex* mobilize more metabolism energy for locomotory activity (Fig. 10).

## V. 3. Cooper effects on other aquatic invertebrates

### V. 3. 1. Acute Toxicity

By studying the scientific literature it can be stated, that no single life stage was consistently most sensitive to copper. Eggs of the snail *Amnicola* sp. were less sensitive than adults (Rehwoldt et al., 1973), whereas crayfish eggs (*Orconectes rusticus*) were most sensitive to copper stress (Hubschman 1967 b). The first instars of the midge larvae *Chironomus tentans* were the life stage most sensitive to copper (Nebeker et al., 1984 b). Toxicity of Cu to the amphipod *Hyalella azteca* was consistent across all age classes (Collyard et al. 1994).

Tab. 11 gives an overview about acute toxicity data determined for different aquatic invertebrates in several studies:

Chapter V – General discussion

Tab. 11: Acute copper toxicity data on different freshwater invertebrates

| Species | Lifestage | Test Type | $LC_{50}$-value [mg Cu l$^{-1}$] | Reference |
|---|---|---|---|---|
| *Asellus meridianus* | 4 – 6 mm | SS | 48-h $LC_{50}$ = 1.2-2.5 | Brown 1976 |
| *Cambarus robustus* | intermoult adult | S | 24/96-h $LC_{50}$ = 0.83 – 3.48 | Taylor et al. 1995 |
| *Campeloma decisum* | 11 – 27 mm | FT, $CuSO_4$ | 96-h $LC_{50}$ = 1.700 | Arthur, Leonard 1970 |
| *Chironomus riparius* | 2$^{nd}$ instar | S | 48-h $LC_{50}$ = 1.170 | Dobbs et al. 1994 |
| *Chironomus tentans* | 2$^{nd}$ instar<br>3$^{rd}$ instar<br>4$^{rd}$ instar | FT, $CuCl_2$ | 96-h $LC_{50}$ = 0.773<br>96-h $LC_{50}$ = 1.446<br>96-h $LC_{50}$ = 1.690 | Nebeker et al. 1984 b |
| *Gammarus pulex* | 3 – 5 mm | SS (1/2 d) | 48-h $LC_{50}$ = 0.037 | Taylor et al. 1991 |
| *Physa integra* | 4 – 7 mm | FT, $CuSO_4$ | 96-h $LC_{50}$ = 0.039 | Arthur and Leonard 1970 |
| *Physella sp.* | < 10 mm | S | 48-h $LC_{50}$ = 0.109 | Dobbs et al. 1994 |

Legend: Test Type: S = Static, SS = Semistatic, FT = Flow Through


V. 3. 2. Physiological and sublethal effects of Cu to aquatic invertebrates

Several ethotoxicological studies showed that Cu reduced the activity of many aquatic invertebrates. Copper influences a variety of physiological parameters to explain these ethotoxicological effects. The most important effect of Cu is on the nervous system of the organisms. S.-Rozsa and Salanki (1990) showed that Cu modified the effect of neurotransmittors and ionic current in the nerve systems of the pond snail *Lymnaea stagnalis*. Other influences could be shown on the osmoregulatory physiology. This effect were shown for the snail *Biomphalaria glabrata* due to Cu exposure (Sullivan and Cheng, 1975). Effects of Cu on the osmotic and ionic regulation were also shown for the crayfish *Oronectes rusticus* (Hubschmann, 1967 a). Here the antennal glands degenerated to varying degrees when exposed to low concentrations of copper. Respiratory enzymes were inhibited at Cu concentrations greater than 1 mg l$^{-1}$. Mersch et al. (1993) showed that the filtration rate of the zebra mussel *Dreissena polymorpha* was inhibited at 0.06 mg l$^{-1}$ Cu. Sublethal levels of Cu (0.25 – 1 ppm) resulted in increased respiration rates in the freshwater mussel *Lamellidens marginalis* compared to controls (Raj and Hameed, 1991).

Chapter V – General discussion

V. 3. 2. 1. Bioconcentration of Cu

In general the fate of a xenobiotic compound in the organism (toxicokinetics) can be described with four stages: Uptake → Dispersion → Metabolism → Excretion. All of these stages are very important for the bioaccumulation and toxicity of the chemical compound in the organism (Fent, 1998).

An overview about toxicokinetic processes of a xenobiotica is shown in Fig. 19.

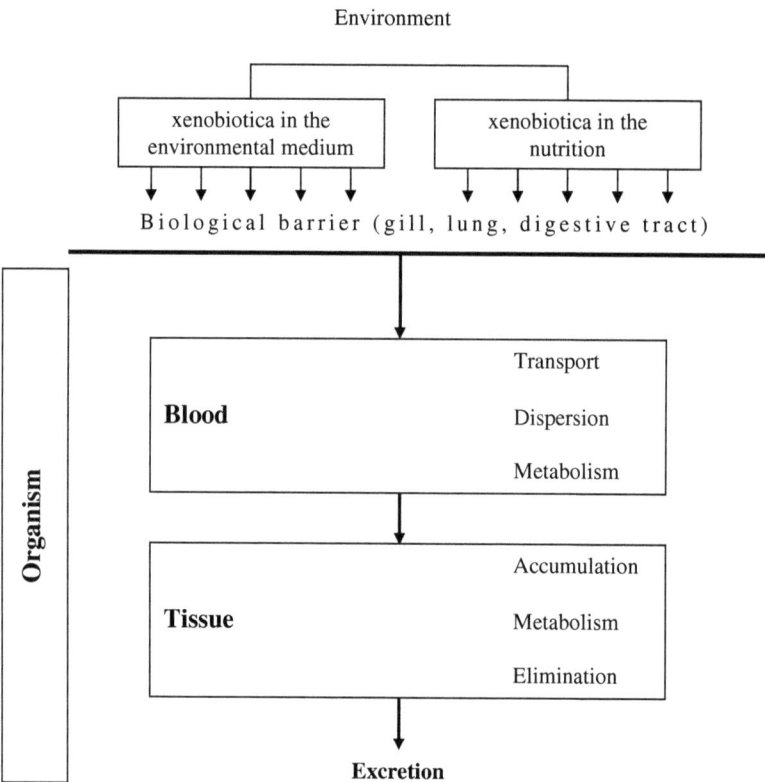

Fig. 19: Uptake and fate of chemical compounds in organism (modified after Fent, 1998)

## Chapter V – General discussion

The effect of animal size on the Cu concentration in animals is not clear. Cu concentrations in the Pacific oyster (*Crassostrea gigas*) increased with size (Ayling, 1974). Borgmann and Norwood (1995) showed that Cu concentrations were independent of the size of the amphipod *Hyalella azteca*. Uptake of Cu was rapid but these amphipods were able to control body burdens gradually and after long exposure periods. Metal concentrations in four species of mayfly larvae decreased with an increase in size (Jop, 1991).

In crayfish, Cu concentrations were highest in the hepatopancreas and gills (*Cambarus bartoni*) (Zia and Alikhan, 1989). Tessier et al. (1984) measured highest copper concentrations in gills and mantle of the freshwater mollusc *Elliptio complanata*.

Uptake of copper by the clam *Anadonta cygnea* was linear and a depuration of Cu from gills did not occur whereas Cu concentrations in foot and mantle decreased in a linear fashion (Salanki and V.-Balogh, 1989). Bioconcentration of copper by daphnids was reduced in water of greater hardness and higher concentration of humic acids, the latter only in older daphnids (Winner, 1985). Uptake of copper by the isopod *Asellus aquaticus* increased at greater temperatures, but pH over a range from 5 to 8 did not significantly affect copper concentrations in the isopod. Since Cu elimination could not be observed it is stated, that Cu was bound in a stable body compartment (Van Hattum et al., 1993).

Copper concentrations in chironomids were positively related to easily reducible Cu sediment concentrations and negatively with easily reducible iron sediment concentration (Young and Harvey, 1991). Cu concentrations in the crayfish (*P. clarkii* and *P. acutus*) were unrelated to sediment copper concentrations (Finerty et al., 1990).

A further process, which is very important for the evaluation of the toxic potential of heavy metals is to be informed about the accumulation over the food web (biomagnification). The metal transfer over the food web depends on the chemical structure in what the metal is stored in the tissues of the animals. For example it could be shown, that heavy metals, like Mn, Ni, Cu, Zn and Ag passes the gut of marine pagurids, which consumed midgut glands of marine molluscs. These metals can be analysed in the feces. Other metals, like Cd and Cr have been absorbed by the tissue of the crustaceans (Nott and Nicolaidou, 1993).

## V. 4. Cadmium effects on crustaceans

### V. 4. 1. Acute toxicity

It is known, that the toxic potential of Cd is enormous, since continuous exposure of organisms to already a low concentration may result in bioaccumulation and biomagnification (Kljaković Gašpic et al., 2002).

An overview about acute toxicity data of Cadmium to different crustacean species is given in Tab. 12.

Tab. 12: Overview about acute toxicity data of Cd to different crustacean species

| Species | Lifestage | $LC_{50}$ [mg $l^{-1}$] | Reference |
|---|---|---|---|
| Hippolyte inermis | 14 mm | 12h-$LC_{50}$ = 3.450 mg $l^{-1}$ | this work |
| Niphargus aquilex | n. d. | 96h-$LC_{50}$ = 6.500 (pH 5.0)<br>96h-$LC_{50}$ = 5.000 (pH 6.0)<br>96h-$LC_{50}$ = 4.500 (pH 7.0)<br>96h-$LC_{50}$ = 2.450 (pH 8.0) | Meinel et al., 1989 |
| Caecidatea bicrenata | n. d. | 96h-$LC_{50}$ = 2.200 (pH 7.0)<br>96h-$LC_{50}$ = 1.200 (pH 7.2) | Bosnak and Morgan, 1981 |
| Caecidatea stygia | n. d. | 96h-$LC_{50}$ = 0.290 (pH 7.5) | Bosnak and Morgan, 1981 |
| Parastenocaris germanica | n. d. | 96h-$LC_{50}$ = 2.200 (pH 6.8) | Notenboom et al., 1992 |
| Asselus aquaticus | Embryo | 96h-$LC_{50}$ = 2.000 µg $l^{-1}$ | |
| Asselus aquaticus | Juvenile, 2.3 mm | 96h-$LC_{50}$ = 1.700 mg $l^{-1}$ | Green et al., 1986 |
| Asselus aquaticus | Adult, 9.87 mm | 96h-$LC_{50}$ = 1.000 mg $l^{-1}$ | |

Legend: n. d.: not described

### V. 4. 2. Physiological and sublethal effects of Cd to aquatic invertebrates

Since Cd is considered to be one of the most toxic metal, toxic effects of this non-essential trace metal are well described in scientific literature (Barata et al., 1998; Wolf et al., 1998; Rasmussen and Andersen, 2000; Salánki et al., 2003; Filipović and Raspor, 2003). Knops et al. (2001) investigated alterations of physiological energetics, growth and reproduction of *Daphnia magna* due to Cd-stress. They found out, that Cd reduced the weight and the body length of daphnids indicating an impaired growth rate.

Chapter V – General discussion

Wolf et al. (1998) explained a decreased swimming activity and – velocity of Cd-stressed *D. magna* by the loss of fitness, due to the effect of the metal on metabolism inducing a loss in energy production. This leaves the organism with less energy for muscle activity or locomotion and overcoming the friction of the aquatic medium during swimming activity. An inhibition of the activity of the enzyme carbonic anhydrase (CA) of the estuarine crab *Chasmagnatus granulata* due to toxic Cd exposure was shown by Vitale et al. (1999). The authors calculated a 96h-$EC_{50}$ value (CA inhibition) from 1,58 ppm. This $EC_{50}$ value was more sensitive than the corresponding 96h-$LC_{50}$ value from 2.69 mg $l^{-1}$.

V. 4. 2. 1. Bioconcentration of Cd

In crustaceans, cadmium uptake from water occurs mainly through the gill epithelium (Rainbow, 1997 a). The most obvious parameter controlling cadmium uptake is its own concentration in water, but there is also a well documented role of salinity: the uptake, accumulation and toxicity of cadmium increase when the salinity of the water decreases (Wildgust and Jones, 1998). The uptake process of Cd depends on two factors, namely the cadmium speciation and physiological processes, as for instance osmoregulation (Rainbow 1997 a).

Bouchet et al. (2000) investigated the bioaccumulation of Cd in the oligochaete *Tubifex tubifex*. Their result showed a rapid Cd bioaccumulation since worms exposed to 0.005 mg $l^{-1}$ Cd accumulated 80.4 µg Cd $g^{-1}$ dry weight, which represented a net uptake of 75.2 µg $g^{-1}$ and a concentration factor of 15.040. Salánki et al. (2003) found out, that molluscs and also other invertebrates and vertebrates living in Lake Balaton accumulate persistent toxic substances like Cd. In their studies they showed, that Cd accumulated in the biomass of crustaceans with an average concentration of 0.1 – 0.5 mg $kg^{-1}$ dry weight. Molluscs investigated in their studies, like *A. cygnea* accumulated Cd mainly in the outer (13.1 ± 2.1 mg $kg^{-1}$ dry weight) and inner gill (13.4 ± 2.1 mg $kg^{-1}$ dry weight) but also in other tissues, like the mantle (14.7 ± 3.2 mg $kg^{-1}$ dry weight) and the foot tissue (9.0 ± 1.6 mg $kg^{-1}$ dry weight). Toxicokinetic studies, using the amphipods *Gammarus zaddachi* and *G. salinus* as testorganisms, showed kinetic bioconcentration factors (BCFs) for Cd in the range of 400 – 1200 with water born Cd exposures ranging from 17 to 52 µg Cd $l^{-1}$ (Clason and Zauke, 2000).

Chapter V – General discussion

## V. 5. Mechanisms of metal detoxification in organisms

It is known, that there exist a number of metal detoxification and resistance mechanisms in organisms. This includes resistance to metals that are purely toxic to the organism and serve no biological function such as mercury and cadmium, and also extends to metals such as copper and zinc that are toxic at high concentrations but absolutely essential in trace amounts (Ybarra and Webb, 1998).

The first mechanism involves extracellular binding. Cells may synthesize and release organic materials that chelate metals and reduce their bioavailability (Clarke et al., 1987) or the metal ions may be bound to the outer cell surface. These complexed forms are generally not readily transported into the cell because of structure and complexity. Secondly, cells can increase the excretion rate of certain metal ions using energy-driven efflux pumps. Internal metal sequestration, a third resistance mechanism, is one of the most important mechanisms by which bacteria combat heavy metal exposure and subsequent accumulation (Ybarra and Webb, 1998).

For aquatic invertebrates three fundamental mechanisms of cation homeostasis have been identified: (i) binding on soluble ligands, like metallothioneins, (ii) binding within membran associated vesicles (e. g. lysosomes) and (iii) formation of unsoluble precipitates (Ritterhoff, Zauke, 1997).

### V. 5. 1. Metallothioneins (MTs)

Since it is known, that MTs occur throughout the animal kingdom and are also found in higher plants, eukaryotic microorganisms, and in some prokaryotes the importance of these proteins in metal metabolism have been intensively discussed in several studies (Margoshes and Vallee, 1957; Viarengo, 1997; Riek et al., 1999).

MTs have been discovered in 1957 when Margoshes and Vallee (1957) identified in equine kidney cortex a cadmium-binding protein responsible for the natural accumulation of cadmium in this tissue. MTs are still the only biological compounds known to naturally contain this metal.

# Chapter V – General discussion

However, it is known, that cadmium is one of several optional metallic components, the others being most commonly copper and zinc.

As already mentioned under IV. 4. Metallothioneins are small cysteine-rich proteins able to bind heavy metal ions with high affinity (Mounaji et al, 2002; Viarengo, 1997). MTs are thought to protect cells and organisms from toxic metal ions and to play an important role in the heavy metal metabolism (Riek et al., 1999). Much of our understanding of the biological actions of the MTs has come from comparative analysis of the chemical and structural features. All MTs have uniquely folded polypeptide structures specifed by the arrangement of the Cys residues in the chain and the coordination preferences of the metal ions.

In this study no explicitly investigations of the MT-content in the cells and tissues of the used test animals have been made. The main focus of this work was to determine behavioural parameters as reliable biological endpoints for the evaluation of toxic heavy metal stress.

However, locomotory behaviour can not be understood without including all relevant physiological mechanisms, which is responsible for this stress behaviour. Seen from this point of view MTs, besides other factors, have to be regarded as one of the main factors in the protection and adaptation reaction of the used organisms.

Most crustaceans have haemocyanin as respiratory pigment in their body fluid. For some decapods it could be observed, that this proteins have been reduced during the molting phase with a subsequent re-displacement of the haemocyanine linked copper in the hepatopancreas, where it will be bound to metallothioneins (Ritterhoff and Zauke, 1997).

## V. 5. Biomonitoring of xenobiotica

Information of the discipline "Biomonitoring" provides us we two primary functions: 1. to control, signal and predict calamities and accidental spills as early warning systems and 2. to document long-term trends, both natural and deleterious (Gerhardt, 1999).

# Chapter V – General discussion

Going through the scientific literature dealing with the field "biomonitoring and ecotoxicology" it can be stated, that the use of organisms for monitoring the pollution of the environment are precious tools of the applied environmental protection.

Biological variables include bioaccumulation , toxicity (responses to chemicals at different biological organisation levels by means of bioassays and biological early warning systems) and ecosystem responses (ecosystem integrity). The evaluation of the bioavailability of a specific chemical compound or xenobiotica cocktail is only possible, when using test organisms as indicators.

As a subdiscipline of Ecotoxicology, the field "Aquatic Toxicology" was born as an illegitime child of classical mammalian toxicology, since the classification of the toxicity of chemicals was primarily evaluated by means of fish tests. Little by little more organisms (including many invertebrate species) has been inegrated into the eco- or ethotoxicological research, respectively.

Since heavy metals are major pollutants in aquatic ecosystems, this work aimed to contribute to a higher understanding of the behaviour of heavy metals in aquatic environments and their (often) negatively effects on the biota. Furthermore the results of this work highlights the sensitivity of such ecosystems and their inhabitants.

As a general conclusion, it can be statet, that behavioural responses of the freshwater cladoceran *D. magna* STRAUS and *D. pulex* LEYDIG and the marine decapod *Hippolyte inermis* LEACH are sensitive bioindicators to heavy metal stress. Changes in the behavioural parameters "swimming velocity" and "- participation" can be used as early stress responses for chronic heavy metal contamination as part of ecological risk assessment.

# VI. References:

Adami, G., Barbieri, P., Fabiani, M., Piselli, S., Predonzani, S., Reisenhofer, E., 2002: Levels of cadmium and zinc in hepatopancreas of reared Mytilus galloprovincialis from the Gulf of Trieste (Italy). Chemosphere 48, 671 – 677.

Alikhan, M. A., Bagatto, G., Shaheen Z, 1990: The crayfish as a "biological indicator" of aquatic contamination by heavy metals. Water Res. 24: 1069-1076.

Anonymus, 1996: Copper. Alberta Water Quality Guideline for the protection of freshwater aquatic life. pp. 123.

Anonymus, 2003: Grundkurs Ökotoxikologie 2003 – Immobilisationstest mit *Daphnia magna*. pp. 6.

Arthur, J. W., Leonard, E. N., 1970: Effects of copper on *Gammarus pseudolimnaeus*, *Physa integra* and *Campeloma decisum* in soft water. J. Fish. Res. Board Can. 27: 1277 – 1283.

Ayling, G. M., 1974: Uptake of cadmium, copper, lead, and chromium in the Pacific oyster, *Crassostrea gigas*, grown in the Tamar River, Tasmania. Water Res. 8: 729 – 738.

Baillieul, M., Scheunders, P. 1998: On-line determination of the velocity of simultaneously moving organisms by image analysis for the detection of sublethal toxicity. Water Research 32 (4), 27-1034.

Barate, C., Baird, D. J., Markich, S. J., 1998: Influence of genetic and environmental factors on the tolerance of *Daphnia magna* Straus to essential and non-essential metals. Aquatic Toxicology 42: 115 – 137.

Beitinger, T. L., McCauley, R. W., 1990: Whole-animal physiological processes for the assessment of stress in fishes. J. Great Lakes Res. 16 (4), 542 – 575.

References

Bishop, W. E., Perry, R. L., 1981: Development and Evaluation of a Flow-Through Growth Inhibition Test with Duckweed (*Lemna minor*). In: Branson D. R. and Dickson K. L. (eds.). Aquatic Toxicology and Hazard Assessment: $4^{th}$. Conference. ASTM STP 737.

Blaylock, B. G., Frank, M. L., McCarthy, J. F., 1985: Comparative toxicity of copper and acridine to fish, *Daphnia* and algae. Eniron. Toxicol. Chem. 4, 63 – 73.

Blübaum-Gronau, E., Hoffmann, M., 1997: Steigerung der Sensitivität eines kontinuierlichen Daphnientestes durch die Berücksichtigung einer Vielzahl von Verhaltensweisen. Vom Wasser 89, 163-173.

Borcherding, J., Volpers, M., 1994: The „Dreissena-Monitor" – First results on the application of this biological early warning system in the continuous monitoring of water quality. Water Science Technology, Vol. 29, No. 3, 199-201.

Borgmann, U., Charlton, C. C., 1984: Copper complexation and toxicity to Daphnia in natural waters. J. Great Lakes Res. 10, 393 – 398.

Borgmann, U., Norwood, W. P., 1995: Kinetics of excess (above background) copper and zinc in *Hyalella azteca* and their relationship to chronic toxicity. Can. J. Fish. Aquat. Sci. 52: 864 – 874.

Bosnak, A. D., Morgan, E. L., 1981: Comparison of acute toxicity for Cd, Cr, and Cu between two distinct populations of aquatic hypogean isopods (*Caecidatea* sp.). $8^{th}$ International Congress of Seleology 1: 72 – 74.

Bouche, M.-L., Habets, F., Biagianti-Risbourg, S., Vernet, G., 2000: Toxic effects and Bioaccumulation of Cadmium in the Aquatic Oligochaete *Tubifex tubifex*. Ecotoxicology and Environmental Safety 46: 246 – 251.

Breuer, H., 2000: dtv-Atlas Chemie – Band 1: Allgemeine und anorganische Chemie. Deutscher Taschenbuch Verlag; 9. Auflage.

# References

Braunbeck, T., Burkhardt-Holm, P., Görge, G., Nagel,R., Negele, D. R., Storch, V., 1992: Regenbogenforelle und Zebrabärbling, zwei Modelle für verlängerte Toxizitätstests: Relative Empfindlichkeit, Art- und Organspezifität in der cytopathologischen Reaktion von Leber und Darm auf Atrazin. Schriftenreihe des Vereins für Wasser-, Boden-, Lufthygiene 89, 109-145.

Braunbeck, T., 1995: Zelltests in der Ökotoxikologie. Landesanstalt für Umweltschutz Baden-Württemberg. Projekt „Angewandte Ökologie" 11.

Brown, B. E., 1976: Observations on the tolerance of the isopod *Asellus meridianus* Rac. to copper and lead. Wat. Res. 10: 555 – 559.

Bulich, A. A., Huynh, H., Ulitzur, S., 1996: The use of luminescent bacteria for measuring chronic toxicity. In: Techniques in Aquatic Toxicology. CRC Press, Inc., 3-12.

Bundesministerium für Umwelt, Naturschutz und Reaktorsicherheit (Hrsg.) (1997): Umweltpolitik – Agenda 21. Konferenz der Vereinten Nationen für Umwelt und Entwicklung im Juni 1992 in Rio de Janeiro.

Cairns, J., Dickson, L. K., Sparks, E. R., Waller, T. W., 1970: A preliminary report on rapid biological information systems for water pollution control. Journal of Water Pollution Control. Fed. 42(5), 685-703.

Charoy, P.C., Janssen, R.C., Persoone, G., Clement, P., 1995: The swimming behaviour of *Brachionus calyciflorus* (rotifer) under toxic stress. I. The use of automated trajectometry for determining sublethal effects of chemicals. Aquatic Toxicology 32, 271 – 282.

Charoy, C., Janssen, R.C., 1999: The Swimming Behaviour of *Brachionus Calyciflorus* (Rotifer) under Toxic Stress. II. Comparative sensitivity of various behavioural criteria. Chemosphere, Vol. 38, No. 14, 3247 – 3260.

Chiaudani, G., Vighi, M., 1978: The use of *Selenastrum capricornutu*m batch cultures in toxicity studies. Mitt. Int. Ver. Limnol. 21, 316-329.

References

Clarke, S.E., et al., 1987: Induction of siderophore activity in Anabaena species and its moderation of copper toxicity. Applied and Environmental Microbiology. 53(5), 917-922.

Clason, B., Zauke, G.-P., 2000: Bioaccumulation of trace metals in marine and estuarine amphipods: Evaluation and verification of toxicokinetic models. Canadian Journal of Fisheries & Aquatic Sciences 57: 1410 – 1422.

Cleveland, L., Fairchild, F. J., Little, E. E., 1999: Biomonitoring and Ecotoxicology: Fish as Indicators of Pollution-Induced Stress in Aquatic Systems. Environmental Science Forum Vol. 96. pp. 195-232. In: Gerhardt, A. (ed.): Biomonitoring of Polluted Water. Trans Tech Publications Ltd. Switzerland

Collyard, S. A., Ankley, G. T., Hoke, R. A., Goldenstein, T., 1994: Influence of age on the relative sensitivity of *Hyalella azteca* to dizinon, alkyl phenol ethoxylates, copper, cadmium and zinc. Arch. Environ. Contam. Toxicol. 26: 110 – 113.

Crease, T.J., Taylor, D.J., 1998: The origin and evolution of variable-region helices in V4 and V7 of the small-subunit ribosomal RNA of branchiopod crustaceans. Molecular Biology and Evolution 15, no. 11, 1430-1446.

Dave, G., 1984: Effects of copper on growth, reproduction, survival and haemoglobin in *Daphnia magna*. Comp. Biochem. Physiol. Vol. 78C, No. 2, 439 – 443.

Demayo, A., Taylor, M. C., 1981: Guidelines for Surface Water Quality. Vol. 1: Inorganic Chemical Substances. Copper. Water Quality Branch, Inland Waters Directorate, Environment Canada, Ottawa.

Dobbs, M. G., Farris, J. L., Reash, R. J., Cherry, D. S., Cairns Jr., J., 1994: Evaluation of the resident-species procedure for developing site-specific water quality criteria for copper in Blaine Creek, Kentucky. Environ. Toxicol. Chem. 13, 963 – 971.

Dodson, S.I., Ramcharan, C., 1991: Size-specific swimming behaviour of *Daphnia pulex*. J. Plankton Res. 13, 1367 – 1379.

# References

Dodson, S.I., D.G. Frey, 1991: Cladocera and Other Branchipoda. In: Ecology and Classification of North American Freshwater Invertebrates. Thorpe, J.H. and A.P. Covich (eds.). Academic Press, Inc. Toronto. Chapter 20, 723-786.

Eberius, M., Vandenhirtz, D., 1999: Einsatz eines speziellen Bildanalysesystems zr ökotoxikologisch umfassenden und kosteneffizienten Auswertung des Wasserlinsentests. In: Oehlmann, J. Markert, B. (eds): Ökotoxikologie – ökosystemare Ansätze und Methoden. ecomed verlagsgesellschaft AG & Co. KG, 167-170.

Englund, M. P. V., Heino, P. M., Melas, G., 1994: Field method for monitoring valve movements of bivalved molluscs. Water Research 28, 2219-2221.

Fent, K., 1998: Ökotoxikologie. Georg Thieme Verlag Stuttgart •New York

Ferrando, M.D., Andreu, E., 1993: Feeding behavior as an index of copper stress in *Daphnia magna* and *Brachionus calyciflorus*. Comp. Biochem. Physiol. Vol. 106 C, No. 2, 327 – 331.

Fichet, D., Radenac, G., Miramand, P., 1998: Experimental Studies of Impacts of Harbour Sediments Resuspension to Marine Invertebrates Larvae: Bioavailability of Cd, Cu, Pb and Zn and Toxicity. Marine Pollution Bulletin, Vol. 36, No. 7 – 12, 509 – 518.

Filipović, V., Raspor, B., 2003: Metallothionein and metal levels in cytosol of liver, kidney and brain in relation to growth parameters of *Mullus surmuletus* and *Liza aurata* from the Eastern Adriatic Sea. Water Research 37: 3253 – 3262.

Finerty, M. W., Madden, J. D., Feagley, S. E., Grodner, R. M., 1990: Effects of environs and seasonality on metal residues in tissues of wild and pond-raised crayfish in southern Louisiana. Arch. Environ. Contam. Toxicol. 19: 94 – 100.

Fomin, A., Oehlmann, J., Markert, B., 2003: Praktikum zur Ökotoxikologie – Grundlagen und Anwendungen biologischer Testverfahren. ecomed verlagsgesellschaft AG & Co. KG. pp. 239.

Fox, R., 2003: Invertebrate Anatomy. Lecture script of the Lander university. pp. 4.

# References

Gambi M.C., Lorenti M., Russo G.F., Scipione M.B., Zupo V., 1992: Depth and seasonal distribution of some groups of vagile fauna of the *Posidonia oceanica* leaf stratum: structural and trophic analyses. PSZN I: Mar Ecol 13,17–39.

Garnacho, E., Peck, L. S., Tyler, P. A., 2001: Effects of copper exposure on the metabolism of the mysid *Praunus flexuosus*. Journal of Experimental Marine Biology and Ecology 265, 181 – 201.

Geller, W., Mäckle, H., 1977: Kontinuierlicher Biotest zur Toxizitätsüberwachung von Trinkwasser. – DVGW-Schriftenreihe 14, 173-181.

Gerhardt, V., Putzger, J., 1992: Ein Biotest zur Gewässerüberwachung auf der Grundlage der verzögerten Fluoreszenz von Algen. Schriftenreihe des Vereins für Wasser-, Boden-, Lufthygiene 89, 277-284.

Gerhardt, A., Clostermann, M., Fridlund, B., Svensson, E. 1994: Monitoring of behvioral pattern of aquatic organisms with an impedance conversion technique. Environment International 20(2), 209-219.

Gerhardt, A., Bisthoven De. J., 1995: Behavioural, developmental and morphological responses of Chironomus gr. thummi larvae (Diptera, Nematocera) to aquatic pollution. Journal of Aquatic Ecosystem Health 4, 205-214.

Gerhardt A., 1995: Monitoring Behavioural Responses to Metals in Gammarus pulex (L.) (Crustacea) with Impedance Conversion. ESPR – Environ. Sci. & Pollut. Res. 2(1), 15-23

Gerhardt, A., Carlsson, A. Ressemann, C., Stich, P. K., 1998: New Online Biomonitroing System for Gammarus pulex (L.) (Crustacea): In Situ Test Below a copper Effluent in South Sweden. Environmental Science & Technology/Vol. 32. No. 1, 150-156.

Gerhardt, A., 1999: Recent Trends in Online Biomonitoring for Water Quality Control. Environmental Science Forum Vol. 96, 95-118. In : Gerhardt, A. (2000) (ed.): Biomonitoring of Polluted Water. Trans Tech Publications Ltd, Switzerland.

References

Gerhardt, A., 2000: Der Multispecies Freshwater Biomonitor (MFB) und seine Anwendungsmöglichkeiten in Forschung und Monitoring von Wasserqualität. Deutsche Gesellschaft für Limnologie (DGL) – Tagungsbericht 1999 (Rostock), Tutzing 2000, 972-977.

Gerhardt, A., Quindt, K., 2000: Abwassertoxizität und –überwachung mit den Bachflohkrebsen Gammarus pulex (L.) und Gammarus tigrinus (Sexton) (Crustacea: Amphipoda). Wasser & Boden, 52/10, 19-26, Blackwell Wissenschafts-Verlag, Berlin.

Green, D. W. J., Williams, K. A., Pascoe, D., 1986: The Acute and Chronic Toxicity of Cadmium to Different Life History Stages of the Freshwater Crustacean *Asellus aquaticus* (L)

Ham, D. K., Peterson, J. M., 1994: Effect of fluctuating low-level chlorine concentrations on valve-movement behavior of the asiatic clam (*Corbicula fluminea*). Environmental Toxicology and Chemistry 13, 493-498.

Heath, A.G., 1995: Water Pollution and Fish Physiology , $2^{nd}$ edn. Lewis Publishers, Boca Raton, FL.

Hendriks, J. A., Stouten A. D. M., 1993: Monitoring the Response of Microcontaminants by Dynamic *Daphnia magna* and *Leuciscus idus* Assays in the Rhine Delta: Biological Early Warning as a Useful Supplement. Ecotoxicology and Environmental Safety 26, 265 – 279.

Herkovits, J., Helguero, L., A., 1998: Copper toxicity and copper-zinc interactions in amphibian embryos. The Science of the Total Environment 221, 1 – 10.

Hoffmann, M., Blübaum-Gronau, E., Krebs, F., 1994: Die Schalenbewegung von Muscheln als Indikator von Schadstoffen in der Gewässerüberwachung. Schriftenreihe des Vereins für Wasser-, Boden- und Lufthygiene 93, 125-149, Gustav Fischer Verlag, Stuttgart.

Hubschmann, J. H., 1967 a: Effects of copper on the crayfish Orconectes rusticus (Giard). II. Mode of toxic action. Crustaceana 12: 141 – 150.

# References

Hubschmann, J. H., 1967 b: Effects of copper on the crayfish Orconectes rusticus (Giard). I. Acute toxicity. Crustaceana 12: 33 – 42.

IRC, 1987: *Rhine Action Program* - International Rhine Committee, Strasburg, France.

Jop, K. M., 1991: Concentration of metals in various larval stages of four Ephemeroptera species. Bull. Environ. Contam. Toxicol. 46: 901 – 905.

Jop, K. M., Askew, A. M., Texeira, D. J., MacGregor, J., 1993: Quality Control in Aquatic Toxicity Testing Programs: Evaluation of Copper and Hexavalent chromium as reference toxicants. In: Landis, W. G., Hughes, J. S., Lewis, M. A., (eds.). Environmental Toxicology and Risk Assessment. ASTM, STN 1179, Philadelphia.

Juhnke, I., Besch, K. W., 1971: Eine neue Testmethode zur Früherkennung akut toxischer Inhaltsstoffe im Wasser. Gewässer und Abwasser 51/52, 107-114.

Klein, B., 1992: Umweltmonitoring mit dem Leuchtbakterientest: das Problem falsch negativer Befunde – Kurzmitteilung. Schriftenreihe des Vereins für Wasser-, Boden-, Lufthygiene 89, 653-655.

Kljaković Gašpić, Z., Zvonarić, T., Vrgoč, N., Odžak, N., Barić, A., 2002: Cadmium and lead in selected tissues of two commercially important fish species from the Adriatic Sea. Water Research 36, 5023 – 5028.

Knie, J., 1978: Der Dynamische Daphnientest – ein automatischer Biomonitor zur Überwachung von Gewässern und Abwässern. Wasser und Boden 12, 310-312.

Knops, M., Altenburger, R., Segner, H., 2001: Alterations of physiological energetics, growth and reproduction of *Daphnia magna* under toxicant stress. Aquatic Toxicology 53, 79 – 90.

Kosalwat, P., Knight, A. W., 1987: Chronic toxicity of Cu to a partial life cycle of *Chironomus decorus*. Arch. Environm. Contam. Toxicol. 16 (3), 283 – 208.

# References

Krebs, F., 1992: Der Leuchtbakterientest für die Wassergesetzgebung. Schriftenreihe des Vereins für Wasser-, Boden-, Lufthygiene 89, 591-624.

Lagadic, L., Caquet, T., Ramade, F., 1994: The role of biomarkers in environmental assessment. 5. Invertebrate populations and communities. Ecotoxicology 3, 193 – 208.

Lewis, M. A., 1983: Effect of loading density on the acute toxicities of surfacants, sopper and phenol to *Daphnia magna* Straus. Arch. Environ. Contam. Toxicol. 12, 51 – 55.

Lett, P. F., Farmer, G. J., Beamish, F. W. H., 1976: Effect of copper on some aspects of the bioenergetics of rainbow trout (*Salmo gairdneri*). J. Fish. Res. Board Can. 33: 1335 – 1342.

Link, M., 1992: Zum physiologischen Hintergrund des Leuchtbakterientests. Schriftenreihe des Vereins für Wasser-, Boden-, Lufthygiene 89, 625-632.

Margoshes, M., Vallee, B.L., 1957: A cadmium protein from equine kidney cortex. J. Am. Chem. Soc. 79, 4813 – 4814.

Mastin, B. J., Rodgers, J., H. Jr., 2000: Toxicity and Bioavailability of Copper Herbicides (Clearigate, Cutrine-Plus, and Copper Sulfate) to Freshwater Animals. Archives of Environmental Contamination and Toxicology, 39, 445 – 451.

Meinel, W., Krause, R., 1988: Zur Korrelation zwischen Zink und verschiedenen pH-Werten in ihrer toxischen Wirkung auf einige Grundwasser-Organismen. Zeitschrift für angewandte Zoologie 75: 159 – 182.

Mersch, J., Morhain, E., Mouvet, C., 1993: Laboratory accumulation and depuration of copper and cadmium in the freshwater mussel *Dreissena polymorpha* and the aquatic moss *Rhynchostegium riparioides*. Chemosphere 8, 1475 – 1485.

Merschhemke, C., Regh, W., 1992: Das FluOx-Meßsystem: Ein automatisches Algentestgerät zur kontinuierlichen Gewässerüberwachung. Schriftenreihe des Vereins für Wasser-, Boden-, Lufthygiene 89, 285-292.

References

Mounajia, K., Erraissa, N. E., Wegnez, M., 2002: Identification of Metallothionein in *Pleurodeles waltl*. Z. Naturforsch. 57c, 727 – 731.

Mule, M. B., Lomte, V. S., 1994: Effect of heavy metals ($CuSO_4$ and $HgCl_2$) on the oxygen consumption of the freshwater snail, *Thiara tuberculata*. J. Env. Biol. 15, 263 – 268.

Nakamura, S.I., Oda, J., Shimada, T., Oki, I., 1987: SOS-inducing acitvity of chemical carcinogens and mutagens in *Salmonella typhimurium* TA 1535/pSK1002: examination with 152 chemicals. Mutation Research 192, 239-246.

Nebeker, A. V., Cairns, M. A., Wise, C. M., 1984 b: Relative sensitivity of *Chironomus tentans* life stages to copper. Environ. Toxicol. Chem. 3: 151 – 158.

Nelson, H., Benoit, D., Erickson, R., Mattson, V., Lindberg, J., 1986: The Effects of Variable Hardness, pH, Alkalinity, Suspended Clay, and Humics on the Chemical Speciation and Aquatic Toxicity of Copper. EPA/600/3-86/023. PB86-1714444. pp 132.

Nendza, M., 1987: Toxizitätsbestimmungen von umweltrelevanten Chemikalien mit einem neuen Biotestsystem, Ermittlung physikochemischer Eigenschaften und Ableitung quantitativer Struktur-Toxizitäts-Beziehungen unter Anwendung von Multiregressions- und Hauptkomponenten-Analyse. Dissertation an der Universität Kiel. pp. 157.

Nentwig, H., 1995: Humanökologie – Fakten, Argumente, Ausblicke. Springer Verlag Berlin Heidelberg New York, pp. 588.

Neururer, H., 1975: Abgeänderter Kressewurzeltest. Zeitschrift für Pflanzenkrankheiten und Pflanzenschutz. 82. Bd, H 5/75.

Nohava, M., 1994: Der Leuchtbakterientest in der Umweltkontrolle. UBA-94-090 – Reports. pp. 50.

Notenboom, J., Cryus, K., Hoekstra, J., van Beelen, P., 1992: Effect of Ambient Oxygen Concentration upon the Acute Toxicity of Chlorophenols and Heavy Metals to the

# References

Groundwater Copepod *Parastenocaris germanica* (Crustacea). Ecotoxicology and Environmental Safety 24: 131 – 143.

Nott, J. A., Nocolaidou, A., 1993: Bioreduction of zinc and manganese along a molluscan food chain. Comp. Biochem. Physiol. A. Comp. Physiol., 104, 235 – 238.

Organisation for Economic Co-operation and Development (OECD) (1994), Risk Reduction Monograph No. 5: Cadmium OECD Environment Directorate, Paris, France.

Oehlmann, J., Markert, B., 1999: Ökotoxikologie – Ökosystemare Ansätze und Methoden. ecomed verlagsgesellschaft AG & Co. KG

Overmeyer, S., Hostert, E., Lindner, E. S., Schnabl, H., Peichl, L., 1994: Der Protoplastenbiotest – ein Wirkungstest zur summarischen Schadstofferfassung in der Umwelt. In: Pluta, J. H., Knie, J. Leschber, R. (eds): Biomonitore in der Gewässerüberwachung. Schriftenreihe des Vereins für Wasser-, Boden-, Lufthygiene 93, 291-299, Gustav Fischer Verlag, Stuttgart.

Pacyna, J. M., Scholtz, M. T., Li Y.-F., 1995: Global budget of trace metal sources. Environmental Review 3, 145 – 159.

Pescheck, R., Herlicska, H., 1990: Schadstoffbelastung von Wasser und Abwasser in Österreich. Umweltbundesamt (Ed.), Monographie Bd. p. 24.

Petry, H., 1989: Automatisiertes Frühwarnsystem zur kontinuierlichen Gewässerkontrolle mit Tubificiden als Schadstoffindikatoren. Z. Wasser- Abwasser- Forsch. 22, 120-124.

Pirow, R., Bäumer, C., Paul, R. J., 2001: Benefits of haemoglobin in the cladoceran crustacean *Daphnia magna*. The Journal of Experimental Biology 204, 3425 – 3441.

Putzger, J., Gerhardt, V., 1994: Regensburger Leuchtbakterientest – Erweiterung des DF-Algentests zum on-line-Monitor mit Bacterium Phosphoreum. Schriftenreihe des Vereins für Wasser-, Boden-, Lufthygiene 93, 205-213.

References

Rainbow, P. S., 1997 a: Ecophysiology of trace metal uptake in crustaceans. Estuar. Coast. Shelf. Sci. 44: 169 – 175.

Rainbow, P. S., Wolowicz, M., Fialkowski, W., Smith, B. D., Sokolowski, A., 2000: Biomonitoring of trace metals in the gulf of Gdansk, using mussels (*Mytilus trossulus*) and barnacles (*Balanus improvisus*). Water Research, Vol. 34, No. 6, 1823 – 1829.

Raj, A. I. M., Hameed, P. S., 1991: Effect of copper, cadmium and mercury on metabolism of the freshwater mussel *Lamellidens marginalis* (Lamarck). J. Environ. Biol. 12: 131 – 135.

Rasmussen, A. D., Andersen, O., 2000: Effects of cadmium exposure on volume regulation in the lugworm, *Arenicola marina*. Aquatic Toxicology 48, 151 – 164.

Rehwoldt, R., Lasko, L., Shaw, C., Wirhowski, E., 1973: The acute toxicity of some heavy metal ions toward benthic organisms. Bull. Environ. Contam. Toxicol. 10: 291 – 294.

Riedl, R., 1983: Fauna und Flora des Mittelmeeres. 3. Auflage. Paul Parey Verlag, pp. 836.

Rinderhagen, M., Ritterhoff, J., Zauke, G.-P., Crustaceans as Bioindicators. In: Gerhardt, A. (2000) (ed.): Biomonitoring of Polluted Water. Trans Tech Publications Ltd, Switzerland, 161- 194.

Ritterhoff, J., Zauke, G.-P., 1997: Untersuchungen zur Eignung mariner Crustaceen für ein Biomonitoring von Schwermetallen am Beispiel von arktischem Zooplankton. Forschungsbericht 102 04 274 (UBA-FB 93-115), pp. 56.

Reiter, C., 1994: Entwicklung eines automatisierten Leuchtbakterientests zur kontinuierlichen Abwasserüberwachung. Dissertation an der Technischen Universität Clausthal. pp. 167.

Riek, R., Prěcheur, B., Wang, Y., Mackay, E. A., Wider, G., Güntert, P., Liu, A., Kägi, J. H. R., Wüthrich, K., 1999: NMR Structure of the Sea Urchin (*Strongylocentrotus purpuratus*) Metallothionein MTA. J. Mol. Biol. 291, 417 – 428.

# References

Rodinger, W., 1994: Biomonitoring – Ein Instrument zur kontinuierlichen Überwachung von Gewässern. ÖWAV-Seminar, 7 – 11.

Rodinger, W., 1997: Ökotoxikologische Prüfungen von Immissionen im Donaueinzugsgebiet. 32. Konferenz der IAD, Wien – Österreich 1997 – Wissenschaftliche Hauptreferate.

Salánki, J., V.-Balogh, K., 1989: Physiological background for using freshwater mussels in monitoring copper and lead pollution. Hydrobiol. 188/189: 445 – 454.

Salánki, J., 1992: Heavy metal induced behaviour modulation in mussels: possible neural correlates. Acta Biologia Hungarica 43 (1 – 4), 375 – 386.

Salánki, J., Farkas, A., Kamardina, T., Rózsa, K. S., 2003: Molluscs in biological monitoring of water quality. Toxicology Letters 140 – 141: 403 – 410.

S.-Rozsa, K., Salanki, S., 1990: Heavy metal regulate physiological and behavioral events by modulating ion channels in neuronal membranes of molluscs. Environ. Monitor. Assessm. 14, 363 – 375.

Schmitz, P., Irmer, U., Krebs, F., 1994: Automatische Biotestverfahren in der Gewässerüberwachung. In: Pluta, H. J., Knie, J., Leschber, R., (Eds.): Biomonitore in der Gewässerüberwachung. Schriftenreihe des Vereins für Wasser-, Boden- und Lufthygiene 93, 1- 17.

Schramm, M. et al., 1999: Cellular, Histological and biochemical Biomarkers. Environmental Science Forum Vol. 96, Trans Tech Publications pp. 33-64.

Schwoerbel, J., 1994: Methoden der Hydrobiologie. Gustav Fischer Verlag Stuttgart Jena, 4. Auflage.

Sekulić, B., Vertačnik, A., 1997: Comparison of anthropological and „natural" input of substances through waters into Adriatic, Baltic and Black sea. Water Research 31, 3178 - 3182.

References

Sell, A. F., 1998: Adaptation to oxygen deficiency: Contrasting patterns of haemoglobin synthesis in two coexisting Daphnia species. Comparative Biochemistry and Physiology Part A 120, 119 – 125.

Slabbert, L. J., Venter A. E., 1999: Biological assays for aquatic toxicity testing. Water Science Technology Vol. 39, No. 10-11, 367-373.

Spear, P. A., Pierce, R. C., 1979: Copper in the Aquatic Environment: Chemistry, Distribution and Toxicology. NRCC Associate Committee on Scientific Criteria for Environmental Quality. National Research Council of Canada. pp 227.

Spieser, H. O., Scholz, W., Blübaum-Gronau, E., Hoffmann, M., Grillitsch, B., Vogel, C., 1994: Das System Behavio-Quant zur Bioindikation anhand des Verhaltens von Fischen und von anderen Aquatischen Organismen. In: Alef, K., Fiedler, H., Hutzinger, O. (eds.): Umweltmonitoring und Bioindikation. Eco-Informa-'94, Band 5, 429-448, Umweltbundesamt Wien.

Spoor, W. A., Neiheisel, I. W., Drummond, R. A., 1971: An electrode chamber for recording respiratory and other movements of free-swimming animals. Trans. Am. Fish. Soc. 100, 22-28.

Stallwitz, E., Häder, P-D., 1994: Effects of heavy metals on motility and gravitactic orientation of the flagellate *Euglena gracilis*. Europ. J. Protistol. 30 (1994), 18-24.

Stein, P., 1992: Bakterienelektroden mit Synechococcus und Escherichia coli – Ein kontinuierliches Testsystem zur Online-Überwachung von Oberflächengewässern. Schriftenreihe des Vereins für Wasser-, Boden-, Lufthygiene 89, 269-276.

Steinberg, C., Klein, J., Brüggemann, R., 1995: Ökotoxikologische Testverfahren – Übersicht über bestehende Testverfahren, Modellierung in der Ökotoxikologie, Empfehlungen für Normung und Forschung. ecomed verlagsgesellschaft AG & Co. KG, Landsberg.

References

Steinhäuser, G. K., 1992: Der Leuchtbakterientest mit selbstgezogenen und flüssig getrockneten Bakterien. Schriftenreihe des Vereins für Wasser-, Boden-, Lufthygiene 89, 633-652.

Stumm, W., Morgan, J. J., 1981: Aquatic Chemistry. An Introduction Emphasizing Chemical Equilibria in Natural Waters. $2^{nd}$ Edition. A Wiley-Interscience Publication, pp. 780

Sullivan, J. T., Cheng, T. C., 1975: Heavy metal toxicity to *Biomphalaria glabrata* (Mollusca: Pulmonata). Ann. N. Y. Acad. Sci. 266: 437 – 444.

Swain, R. W., Wilson, M. R., Neri, P. R., Porter, S. G., 1977: A new technique for remote monitoring of activity of freshwater invertebrates with special reference to oxygen consumption by naids of Anax sp. and Somatochlora sp. (Odonata). Can. Entomol. 109, 1-8.

Tankere, S. P. C., Statham, P. J., 1996: Distribution of Dissolved Cd, Cu, Ni and Zn in the Adriatic Sea. Marine Pollution Bulletin, Vol. 32, 623 – 630.

Taylor, E. J., Maund, S. J., Pascoe, D., 1991: Toxicity of four common pollutants to the freshwater macroinvertebrates *Chironomus riparius* Meigen (Insecta: Diptera) and *Gammarus pulex* (L.) (Crustacea: Amphipoda). Arch. Environ. Contam. Toxicol. 21: 371 – 376.

Taylor, R. M., Watson, G. D., Alikhan, M. A., 1995: Comparative sublethal and lethal acute toxicity of copper to the freshwater crayfish, Cambarus robustus (Cambaridae, Decapoda, Crustacea) from an acidic metal-contaminated lake and a circumneutral uncontaminated stream. Wat. Res. 29: 401 – 408.

Tessier, A., Campbell, P. G. C., Auclair, J. C., Bisson, M., 1984: Relationships between the partitioning of trace metals in sediments and their accumulation in the tissues of the freshwater mollusc *Elliptio complanata* in a mining area. Can. J. Fish. Aquat. Sci. 41: 1463 – 1472.

Thomas, M., Florion, A., Chretien, D., Terver, D., 1996: Real-time biomonitoring of water contamination by cyanide based on analysis of the continuous electrical signal emitted by the tropical fish, *Apteronotus albifrons*. Water Research 30, 3083-3091.

References

UBA, 1994: Kontinuierliche Biotestverfahren zur Überwachung des Rheins. UBA-Texte erstellt von der Bund/Länder-Projektgruppe „Wirkungstests Rhein", 2-27

UNCED, 1992: Dokumente der Konferenz der Vereinten Nationen für Umwelt und Entwicklung –Dokumente- Agenda 21, Übersetzung und ed. durch das Bundesministerium für Umwelt, Naturschutz und Reaktorsicherheit Bonn.

Untersteiner, H., Kahapka, J., Kaiser, H., 2003: Behavioural response of the cladoceran *Daphnia magna* Straus to sublethal Copper stress – validation by image analysis. Aquatic Toxicology 65, 435 – 442.

van den Hurk, P., Faisal, M., Roberts Jr, M. H., 1998: Interaction of Cadmium and Benzo[a]pyrene in Mummichog (*Fundulus heteroclitus*): Effects on Acute Mortality. Marine Environmental Research, Vol. 46, No. 1 – 5, 525 – 528.

Van Hattum, B., Korthals, G., Van Straalen, N. M., Govers, H. A. J., Joosse, E. N. G., 1993: Accumulation patterns of trace metals in freshwater isopods in sediment bioassays – influence of substrate characteristics, temperature and pH. Water Res. 27: 669 – 684.

Viarengo, A., Ponzano, E., Dondero, F., Fabbri, R., 1997: A Simple Spectrophotometric Method for Metallothionein Evaluation in Marine Organisms: an Application to Mediterranean and Antarctic Molluscs. Marine Environmental Research, Vol. 44, 1. 69 – 84.

Waiwood, K. G., Beamish, F. W. H., 1978: The effect of copper, hardness and pH on the growth of rainbow trout, *Salmo gairdneri*. J. Fish. Biol. 13: 591 – 598.

World Health Organisation (WHO), 1992: Environmental Health Criteria 134 - Cadmium International Programme on Chemical Safety (IPCS) Monograph.

Wiggins, P. R., Frappell, P. B., 2000: The influence of Haemoglobin on Behavioural Thermoregulation and Oxygen Consumption in *Daphnia carinata*. Physiological and Biochemical Zoology 73 (2), 153 – 160.

# References

Wildgust, M. A., Jones, M. B., 1998: Salinity change and the toxicity of free cadmium ion [$Cd^{2+}$] to *Neomysis integer* (Crustacea: Mysidacea). Aquatic Toxicology 41: 187 – 192.

Winner, R. W., 1985: Bioaccumulation and toxicity of copper as affected by interactions between humic acid and water hardness. Water Res. 19: 449 – 455.

Winner, R. W., Taylor, E. W., 1993: The physiological responses of freshwater rainbow trout, Oncorhynchus mykiss, during acutely lethal copper exposure. J. Comp. Physiol. B. 163, 38 – 47.

WIR (1995): Kontinuierliche Biotestverfahren zur Überwachung des Rheins / erarbeitet von der Bund-Länder-Projektgruppe „Wirkungstests Rhein" (WIR). (ed. Umweltbundesamt Berlin), Erich Schmidt Verlag GmbH & Co., Berlin 1995, pp. 285.

Wolf, G., Scheunders, P., Selens, M., 1998: Evaluation of the swimming activity of *Daphnia magna* by image analysis after administration of sublethal Cadmium concentrations. Comparative Biochemistry and Physiology Part A 120, 99 – 105.

Wratten, S. D., 1994: Video Techniques in Animal Ecology and Behaviour. Chapman & Hall, pp. 211.

Ybarra, G. R., Webb, R., 1998: Differential Responses of Groel and Metallothionein genes to divalent metal cations and the oxyanions of arsenic in the cyanobacterium *Synechococcus sp.* Strain PCC7942. Proceedings of the 1998 Conference on Hazardous Waste Research; 76 – 86.

Young, L. B., Harvey, H. H., 1991: Metal concentration in chironomids in relation to the geochemical characteristics of surficial sediments. Arch. Environ. Contam. Toxicol. 21: 201 – 211.

Zago, C., Capodaglio, G., Ceradini, S., Ciceri, G., Abelmoschi, L., Soggia, F., Cescon, P., Scarponi, G., 2000: Benthic fluxes of cadmium, lead, copper and nitrogen species in the northern Adriatic Sea in front of the River Po outflow, Italy. The Science of the Total Environment 246, 121 – 137.

References

Zia, S., Alikhan, M. A., 1989: Copper uptake and regulation in a copper-tolerant decapod *Cambarus bartoni* (Fabricius) (Decapoda, Crustacea). Bull. Environ. Contam. Toxicol. 42: 103 – 110.

Zupo, V., 2000: Effect of microalgal food on the sex reversal of *Hippolyte inermis* (Crustacea: Decapoda). Mar Ecol Prog Ser 201, 251–259.

# VII. Attachments

## VII. 1. Figures

### VII. 1. 1. Testmaterials

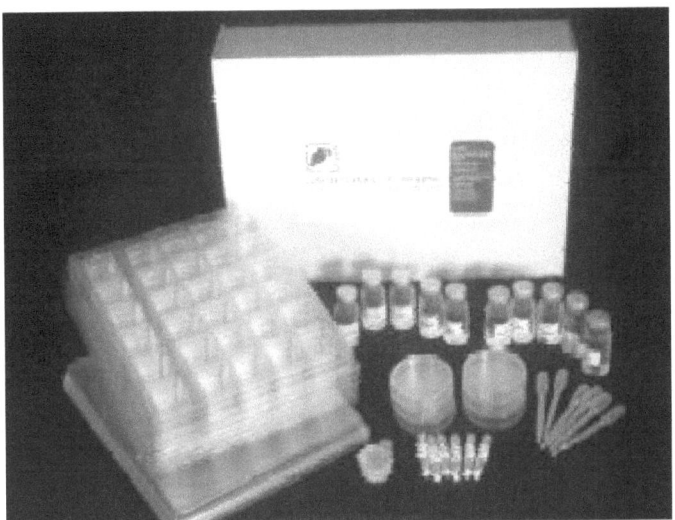

**Fig. 20:** Daphtoxkit set used for the acute and chronic toxicity test (from Microbiotest Inc.) The daphniids (*Daphnia magna* and *D. pulex*) have been delivered in dormant forms (ephippia)

### VII. 1. 2. Test organisms
### VII. 1. 2. 1. Daphniids

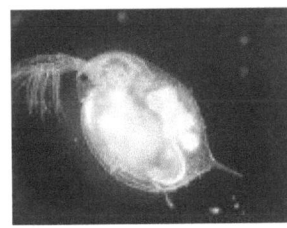  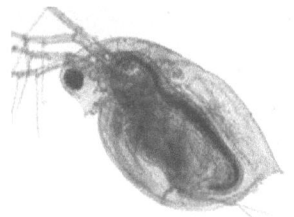

**Fig. 21:** *Daphnia magna* STRAUS          *Daphnia pulex* LEYDIG

## Attachment

### VII. 1. 2. 1. 1. Ephippia

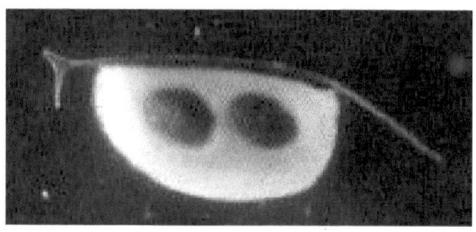
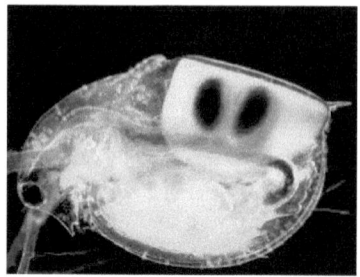

**Fig. 22:** Ephippium from *D. magna* STRAUS     *D. pulex* LEYDIG female with Ephippium

### VII. 1. 2. 2. *Hippolyte inermis* Leach

**Fig. 23:** *Hippolyte inermis* LEACH, seen from (a) lateral, (b) dorsal

Attachment

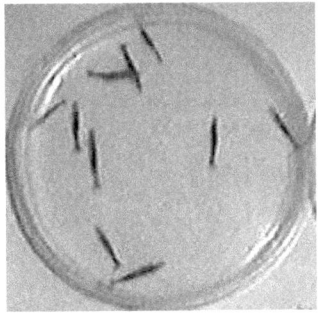

**Fig. 24:** 10 seagrass shrimps in a petri dish, prepared for video-recording

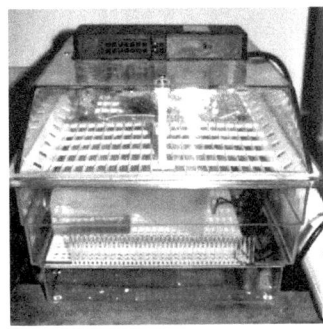

**Fig. 25:** Toxkit-Incubator for keeping temperatures constant during the acute toxicity tests.

**Fig. 26:** Constructed handdredge for catching the test animals

## VII. 2. Statistical data

### VII. 2. 1. Tables and Figures

#### VII. 2. 1. 1. *Hippolyte inermis* Leach

```
* * * * * * * * * * *   P R O B I T     A N A L Y S I S   * * * * * * * * *

Parameter estimates converged after 15 iterations.
Optimal solution found.

Parameter Estimates (PROBIT model:  (PROBIT(p)) = Intercept + BX):

            Regression Coeff.   Standard Error     Coeff./S.E.

 T                 ,58491            ,13874           4,21598


               Intercept     Standard Error    Intercept/S.E.

                -2,01998          ,44707          -4,51830

 Pearson  Goodness-of-Fit  Chi Square =      ,458    DF = 4    P =  ,977

Since Goodness-of-Fit Chi square is NOT significant, no heterogeneity
factor is used in the calculation of confidence limits.

Observed and Expected Frequencies

            Number of   Observed    Expected
     T      Subjects    Responses   Responses    Residual     Prob

      ,00     20,0          ,0         ,434       -,434      ,02169
     1,00     20,0         2,0        1,512        ,488      ,07563
     1,80     20,0         4,0        3,334        ,666      ,16674
     3,20     20,0         8,0        8,822       -,822      ,44107
     5,60     20,0        18,0       17,908        ,092      ,89535
    10,00     20,0        20,0       19,998        ,002      ,99994
```

Attachment

Confidence Limits for Effective T

| Prob | T | 95% Confidence Limits | |
|---|---|---|---|
| | | Lower | Upper |
| ,01 | -,52378 | -3,56095 | ,70574 |
| ,02 | -,05773 | -2,71794 | 1,05183 |
| ,03 | ,23797 | -2,18748 | 1,27582 |
| ,04 | ,46041 | -1,79140 | 1,44728 |
| ,05 | ,64135 | -1,47152 | 1,58906 |
| ,06 | ,79535 | -1,20120 | 1,71168 |
| ,07 | ,93039 | -,96589 | 1,82090 |
| ,08 | 1,05129 | -,75675 | 1,92024 |
| ,09 | 1,16125 | -,56797 | 2,01201 |
| ,10 | 1,26247 | -,39555 | 2,09784 |
| ,15 | 1,68154 | ,30122 | 2,47030 |
| ,20 | 2,01460 | ,82888 | 2,79244 |
| ,25 | 2,30034 | 1,25652 | 3,09384 |
| ,30 | 2,55694 | 1,61644 | 3,38862 |
| ,35 | 2,79472 | 1,92722 | 3,68453 |
| ,40 | 3,02035 | 2,20132 | 3,98611 |
| ,45 | 3,23865 | 2,44809 | 4,29632 |
| **,50** | **3,45349** | **2,67498** | **4,61758** |
| ,55 | 3,66833 | 2,88820 | 4,95251 |
| ,60 | 3,88662 | 3,09318 | 5,30450 |
| ,65 | 4,11225 | 3,29499 | 5,67837 |
| ,70 | 4,35003 | 3,49883 | 6,08121 |
| ,75 | 4,60664 | 3,71082 | 6,52393 |
| ,80 | 4,89237 | 3,93935 | 7,02445 |
| ,85 | 5,22544 | 4,19815 | 7,61543 |
| ,90 | 5,64451 | 4,51534 | 8,36748 |
| ,91 | 5,74572 | 4,59085 | 8,55022 |
| ,92 | 5,85568 | 4,67248 | 8,74915 |
| ,93 | 5,97659 | 4,76178 | 8,96833 |
| ,94 | 6,11162 | 4,86102 | 9,21362 |
| ,95 | 6,26563 | 4,97362 | 9,49395 |
| ,96 | 6,44657 | 5,10520 | 9,82403 |
| ,97 | 6,66901 | 5,26605 | 10,23072 |
| ,98 | 6,96470 | 5,47852 | 10,77270 |
| ,99 | 7,43075 | 5,81088 | 11,62945 |

**Tab. 13:** Probit analysis for the calculation of the 12h-$LC_{50}$-value for Cd-stressed seagrass shrimps. The value for the $LC_{50}$ is bold marked.

Attachment

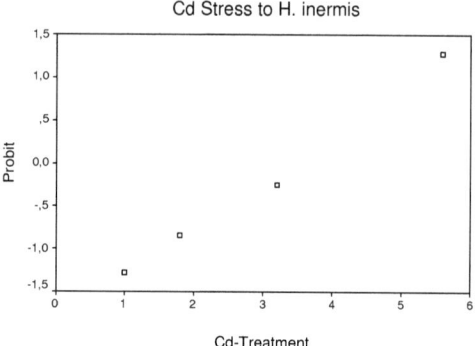

**Fig. 27:** Probit Transformed Responses

| ET | T | velocity [mm s⁻¹] | | | | |
|---|---|---|---|---|---|---|
| | | Central Tendency | | | Dispersion | |
| | | Mean | Median | S. D. | Min | Max |
| Trial start (0 h) | C | 2.14 | 0.63 | 4.11 | 0.00 | 13.45 |
| | $C_1$ | 0.49 | 0.21 | 0.68 | 0.04 | 2.23 |
| | $C_2$ | 0.59 | 0.39 | 0.69 | 0.00 | 1.89 |
| | $C_3$ | 0.18 | 0.00 | 0.54 | 0.00 | 1.71 |
| | | distance moved [mm] | | | | |
| | C | 257.22 | 75.34 | 492.50 | 0.00 | 1613.71 |
| | $C_1$ | 58.18 | 24.86 | 81.61 | 4.59 | 267.58 |
| | $C_2$ | 70.24 | 45.65 | 82.95 | 0.00 | 226.27 |
| | $C_3$ | 22.17 | 0.00 | 64.56 | 0.00 | 205.53 |
| | | velocity [mm s⁻¹] | | | | |
| 1 h | C | 0.74 | 0.34 | 0.82 | 0.00 | 2.04 |
| | $C_1$ | 0.24 | 0.24 | 0.20 | 0.00 | 0.59 |
| | $C_2$ | 0.31 | 0.00 | 0.89 | 0.00 | 2.83 |
| | $C_3$ | 0.03 | 0.00 | 0.07 | 0.00 | 0.23 |
| | | distance moved [mm] | | | | |
| | C | 89.07 | 40.04 | 0.82 | 0.00 | 2.04 |
| | $C_1$ | 27.46 | 20.66 | 24.30 | 0.00 | 70.38 |
| | $C_2$ | 36.71 | 0.00 | 106.68 | 0.00 | 339.32 |
| | $C_3$ | 3.37 | 0.00 | 8.67 | 0.00 | 27.71 |
| | | velocity [mm s⁻¹] | | | | |
| 3 h | C | 1.22 | 0.25 | 1.75 | 0.00 | 4.45 |
| | $C_1$ | 0.06 | 0.05 | 0.05 | 0.00 | 0.23 |
| | $C_2$ | 0.42 | 0.24 | 0.58 | 0.00 | 1.81 |
| | $C_3$ | 0.00 | 0.00 | 0.00 | 0.00 | 0.00 |
| | | distance moved [mm] | | | | |
| | C | 145.70 | 29.58 | 209.86 | 0.00 | 534.14 |
| | $C_1$ | 6.78 | 5.44 | 5.92 | 0.00 | 16.66 |
| | $C_2$ | 50.80 | 28.99 | 69.35 | 0.00 | 217.26 |
| | $C_3$ | 0.00 | 0.00 | 0.00 | 0.00 | 0.00 |

**Tab. 14:** Data of locomotory parameters of *H. inermis* Leach measured for different Cd-concentrations and for different times of Cd-Exposure. N = 10 (per group and time).

Attachment

**Descriptive Statistics**

|  | N | Mean | Std. Deviation | Minimum | Maximum |
|---|---|---|---|---|---|
| velocity [mm/s] | 40 | ,8497 | 2,1833 | ,00 | 13,45 |
| distance moved | 40 | 101,9512 | 261,9597 | ,00 | 1613,71 |
| Treatment | 40 | 1,50 | 1,13 | 0 | 3 |

## Kruskal-Wallis Test

**Ranks**

|  | Treatment | N | Mean Rank |
|---|---|---|---|
| velocity [mm/s] | C | 10 | 27,15 |
|  | C 1 | 10 | 22,55 |
|  | C2 | 10 | 21,55 |
|  | C3 | 10 | 10,75 |
|  | Total | 40 |  |
| distance moved | C | 10 | 27,15 |
|  | C 1 | 10 | 22,55 |
|  | C2 | 10 | 21,55 |
|  | C3 | 10 | 10,75 |
|  | Total | 40 |  |

**Test Statistics[a,b]**

|  | velocity [mm/s] | distance moved |
|---|---|---|
| Chi-Square | 10,814 | 10,814 |
| df | 3 | 3 |
| Asymp. Sig. | ,013 | ,013 |

a. Kruskal Wallis Test
b. Grouping Variable: Treatment

**Tab. 15:** Kruskal-Wallis Test for H. inermis at the beginning of the trial (0h). * $P < 0.05$

## Mann-Whitney Test

**Ranks**

|  | Treatment | N | Mean Rank | Sum of Ranks |
|---|---|---|---|---|
| velocity [mm/s] | C | 10 | 12,15 | 121,50 |
|  | C 1 | 10 | 8,85 | 88,50 |
|  | Total | 20 |  |  |
| distance moved | C | 10 | 12,15 | 121,50 |
|  | C 1 | 10 | 8,85 | 88,50 |
|  | Total | 20 |  |  |

**Test Statistics[b]**

|  | velocity [mm/s] | distance moved |
|---|---|---|
| Mann-Whitney U | 33,500 | 33,500 |
| Wilcoxon W | 88,500 | 88,500 |
| Z | -1,252 | -1,252 |
| Asymp. Sig. (2-tailed) | ,211 | ,211 |
| Exact Sig. [2*(1-tailed Sig.)] | ,218[a] | ,218[a] |

a. Not corrected for ties.
b. Grouping Variable: Treatment

Tab. 16: U-Test comparing the groups C and C1 at the beginning of the trial (0 h). n. s. $P > 0.05$

Attachment

**Ranks**

| | Treatment | N | Mean Rank | Sum of Ranks |
|---|---|---|---|---|
| velocity [mm/s] | C | 10 | 11,95 | 119,50 |
| | C2 | 10 | 9,05 | 90,50 |
| | Total | 20 | | |
| distance moved | C | 10 | 11,95 | 119,50 |
| | C2 | 10 | 9,05 | 90,50 |
| | Total | 20 | | |

**Test Statistics[b]**

| | velocity [mm/s] | distance moved |
|---|---|---|
| Mann-Whitney U | 35,500 | 35,500 |
| Wilcoxon W | 90,500 | 90,500 |
| Z | -1,100 | -1,100 |
| Asymp. Sig. (2-tailed) | ,271 | ,271 |
| Exact Sig. [2*(1-tailed Sig.)] | ,280[a] | ,280[a] |

a. Not corrected for ties.
b. Grouping Variable: Treatment

**Tab. 17:** U-Test comparing the groups C and C2 at the beginning of the trial (0 h). n. s. $P > 0.05$

**Ranks**

| | Treatment | N | Mean Rank | Sum of Ranks |
|---|---|---|---|---|
| velocity [mm/s] | C | 10 | 14,05 | 140,50 |
| | C3 | 10 | 6,95 | 69,50 |
| | Total | 20 | | |
| distance moved | C | 10 | 14,05 | 140,50 |
| | C3 | 10 | 6,95 | 69,50 |
| | Total | 20 | | |

**Test Statistics[b]**

| | velocity [mm/s] | distance moved |
|---|---|---|
| Mann-Whitney U | 14,500 | 14,500 |
| Wilcoxon W | 69,500 | 69,500 |
| Z | -2,773 | -2,773 |
| Asymp. Sig. (2-tailed) | ,006 | ,006 |
| Exact Sig. [2*(1-tailed Sig.)] | ,005[a] | ,005[a] |

a. Not corrected for ties.
b. Grouping Variable: Treatment

**Tab. 18:** U-Test comparing the groups C and C3 at the beginning of the trial (0 h). **$P < 0.01$

**Descriptive Statistics**

| | N | Mean | Std. Deviation | Minimum | Maximum |
|---|---|---|---|---|---|
| velocity [mm/s] | 40 | ,3295 | ,6471 | ,00 | 2,83 |
| distance moved | 40 | 39,1528 | 77,6896 | ,00 | 339,32 |
| Treatment | 40 | 1,50 | 1,13 | 0 | 3 |

Attachment

## Kruskal-Wallis Test

**Ranks**

| | Treatment | N | Mean Rank |
|---|---|---|---|
| velocity [mm/s] | C | 10 | 29,10 |
| | C 1 | 10 | 26,15 |
| | C2 | 10 | 13,90 |
| | C3 | 10 | 12,85 |
| | Total | 40 | |
| distance moved | C | 10 | 29,30 |
| | C 1 | 10 | 25,90 |
| | C2 | 10 | 14,00 |
| | C3 | 10 | 12,80 |
| | Total | 40 | |

**Test Statistics[a,b]**

| | velocity [mm/s] | distance moved |
|---|---|---|
| Chi-Square | 16,490 | 16,494 |
| df | 3 | 3 |
| Asymp. Sig. | ,001 | ,001 |

a. Kruskal Wallis Test
b. Grouping Variable: Treatment

**Tab. 19:** Kruskal-Wallis Test for H. inermis after 1h Cd-exposure. *** $P < 0.001$

## Mann-Whitney Test

**Ranks**

| | Treatment | N | Mean Rank | Sum of Ranks |
|---|---|---|---|---|
| velocity [mm/s] | C | 10 | 11,85 | 118,50 |
| | C 1 | 10 | 9,15 | 91,50 |
| | Total | 20 | | |
| distance moved | C | 10 | 11,95 | 119,50 |
| | C 1 | 10 | 9,05 | 90,50 |
| | Total | 20 | | |

**Test Statistics[b]**

| | velocity [mm/s] | distance moved |
|---|---|---|
| Mann-Whitney U | 36,500 | 35,500 |
| Wilcoxon W | 91,500 | 90,500 |
| Z | -1,022 | -1,097 |
| Asymp. Sig. (2-tailed) | ,307 | ,273 |
| Exact Sig. [2*(1-tailed Sig.)] | ,315[a] | ,280[a] |

a. Not corrected for ties.
b. Grouping Variable: Treatment

**Tab. 20:** U-Test comparing the groups C and C1 after 1h of Cd-Exposure. $P > 0.05$

Attachment

**Ranks**

| | Treatment | N | Mean Rank | Sum of Ranks |
|---|---|---|---|---|
| velocity [mm/s] | C | 10 | 13,75 | 137,50 |
| | C2 | 10 | 7,25 | 72,50 |
| | Total | 20 | | |
| distance moved | C | 10 | 13,80 | 138,00 |
| | C2 | 10 | 7,20 | 72,00 |
| | Total | 20 | | |

**Test Statistics[b]**

| | velocity [mm/s] | distance moved |
|---|---|---|
| Mann-Whitney U | 17,500 | 17,000 |
| Wilcoxon W | 72,500 | 72,000 |
| Z | -2,577 | -2,615 |
| Asymp. Sig. (2-tailed) | ,010 | ,009 |
| Exact Sig. [2*(1-tailed Sig.)] | ,011[a] | ,011[a] |

a. Not corrected for ties.
b. Grouping Variable: Treatment

**Tab. 21:** U-Test comparing the groups C and C2 after 1h of Cd-Exposure. * $P < 0.05$

**Ranks**

| | Treatment | N | Mean Rank | Sum of Ranks |
|---|---|---|---|---|
| velocity [mm/s] | C | 10 | 14,50 | 145,00 |
| | C3 | 10 | 6,50 | 65,00 |
| | Total | 20 | | |
| distance moved | C | 10 | 14,55 | 145,50 |
| | C3 | 10 | 6,45 | 64,50 |
| | Total | 20 | | |

**Test Statistics[b]**

| | velocity [mm/s] | distance moved |
|---|---|---|
| Mann-Whitney U | 10,000 | 9,500 |
| Wilcoxon W | 65,000 | 64,500 |
| Z | -3,125 | -3,163 |
| Asymp. Sig. (2-tailed) | ,002 | ,002 |
| Exact Sig. [2*(1-tailed Sig.)] | ,002[a] | ,001[a] |

a. Not corrected for ties.
b. Grouping Variable: Treatment

**Tab. 22:** U-Test comparing the groups C and C3 after 1h of Cd-Exposure. ** $P < 0.01$

Attachment

**Descriptive Statistics**

|  | N | Mean | Std. Deviation | Minimum | Maximum |
|---|---|---|---|---|---|
| velocity [mm/s] | 40 | ,4765 | 1,1019 | ,00 | 5,44 |
| distance moved | 40 | 57,1937 | 132,1900 | ,00 | 652,63 |
| Treatment | 40 | 1,50 | 1,13 | 0 | 3 |

## Kruskal-Wallis Test

**Ranks**

|  | Treatment | N | Mean Rank |
|---|---|---|---|
| velocity [mm/s] | C | 10 | 22,30 |
|  | C 1 | 10 | 20,80 |
|  | C2 | 10 | 24,20 |
|  | C3 | 10 | 14,70 |
|  | Total | 40 |  |
| distance moved | C | 10 | 22,30 |
|  | C 1 | 10 | 20,85 |
|  | C2 | 10 | 24,15 |
|  | C3 | 10 | 14,70 |
|  | Total | 40 |  |

**Test Statistics[a,b]**

|  | velocity [mm/s] | distance moved |
|---|---|---|
| Chi-Square | 4,237 | 4,208 |
| df | 3 | 3 |
| Asymp. Sig. | ,237 | ,240 |

a. Kruskal Wallis Test
b. Grouping Variable: Treatment

**Tab. 23:** Kruskal-Wallis Test for H. inermis after 2h Cd-exposure. $P > 0.05$

**Descriptive Statistics**

|  |  |  |  |  |  | Percentiles | | |
|---|---|---|---|---|---|---|---|---|
|  | N | Mean | Std. Deviation | Minimum | Maximum | 25th | 50th (Median) | 75th |
| velocity [mm/s] | 40 | ,4235 | 1,0125 | ,00 | 4,45 | ,0000 | 4,500E-02 | ,2500 |
| distance moved | 40 | 50,8195 | 121,4474 | ,00 | 534,14 | ,0000 | 5,4400 | 30,0475 |
| Treatment | 40 | 1,50 | 1,13 | 0 | 3 | ,25 | 1,50 | 2,75 |

# Attachment

## Kruskal-Wallis Test

**Ranks**

| | Treatment | N | Mean Rank |
|---|---|---|---|
| velocity [mm/s] | C | 10 | 29,55 |
| | C 1 | 10 | 19,65 |
| | C2 | 10 | 23,80 |
| | C3 | 10 | 9,00 |
| | Total | 40 | |
| distance moved | C | 10 | 29,60 |
| | C 1 | 10 | 19,60 |
| | C2 | 10 | 23,80 |
| | C3 | 10 | 9,00 |
| | Total | 40 | |

**Test Statistics[a,b]**

| | velocity [mm/s] | distance moved |
|---|---|---|
| Chi-Square | 17,900 | 17,977 |
| df | 3 | 3 |
| Asymp. Sig. | ,000 | ,000 |

a. Kruskal Wallis Test
b. Grouping Variable: Treatment

**Tab. 24:** Kruskal-Wallis Test for H. inermis after 3h Cd-exposure. *** $P < 0.001$

## Mann-Whitney Test

**Ranks**

| | Treatment | N | Mean Rank | Sum of Ranks |
|---|---|---|---|---|
| velocity [mm/s] | C | 10 | 13,95 | 139,50 |
| | C 1 | 10 | 7,05 | 70,50 |
| | Total | 20 | | |
| distance moved | C | 10 | 14,00 | 140,00 |
| | C 1 | 10 | 7,00 | 70,00 |
| | Total | 20 | | |

**Test Statistics[b]**

| | velocity [mm/s] | distance moved |
|---|---|---|
| Mann-Whitney U | 15,500 | 15,000 |
| Wilcoxon W | 70,500 | 70,000 |
| Z | -2,618 | -2,655 |
| Asymp. Sig. (2-tailed) | ,009 | ,008 |
| Exact Sig. [2*(1-tailed Sig.)] | ,007[a] | ,007[a] |

a. Not corrected for ties.
b. Grouping Variable: Treatment

**Tab. 25:** U-Test comparing the groups C and C1 after 3h of Cd-Exposure. ** $P < 0.001$

Attachment

**Ranks**

| | Treatment | N | Mean Rank | Sum of Ranks |
|---|---|---|---|---|
| velocity [mm/s] | C | 10 | 11,60 | 116,00 |
| | C2 | 10 | 9,40 | 94,00 |
| | Total | 20 | | |
| distance moved | C | 10 | 11,60 | 116,00 |
| | C2 | 10 | 9,40 | 94,00 |
| | Total | 20 | | |

**Test Statistics[b]**

| | velocity [mm/s] | distance moved |
|---|---|---|
| Mann-Whitney U | 39,000 | 39,000 |
| Wilcoxon W | 94,000 | 94,000 |
| Z | -,838 | -,838 |
| Asymp. Sig. (2-tailed) | ,402 | ,402 |
| Exact Sig. [2*(1-tailed Sig.)] | ,436[a] | ,436[a] |

a. Not corrected for ties.
b. Grouping Variable: Treatment

**Tab. 26:** U-Test comparing the groups C and C2 after 3h of Cd-Exposure. $P > 0.05$

**Ranks**

| | Treatment | N | Mean Rank | Sum of Ranks |
|---|---|---|---|---|
| velocity [mm/s] | C | 10 | 15,00 | 150,00 |
| | C3 | 10 | 6,00 | 60,00 |
| | Total | 20 | | |
| distance moved | C | 10 | 15,00 | 150,00 |
| | C3 | 10 | 6,00 | 60,00 |
| | Total | 20 | | |

**Test Statistics[b]**

| | velocity [mm/s] | distance moved |
|---|---|---|
| Mann-Whitney U | 5,000 | 5,000 |
| Wilcoxon W | 60,000 | 60,000 |
| Z | -3,724 | -3,724 |
| Asymp. Sig. (2-tailed) | ,000 | ,000 |
| Exact Sig. [2*(1-tailed Sig.)] | ,000[a] | ,000[a] |

a. Not corrected for tios.
b. Grouping Variable: Treatment

**Tab. 27:** U-Test comparing the groups C and C3 after 3h of Cd-Exposure. *** $P < 0.001$

## VII. 2. 1. 2. *Daphnia magna* Straus

| N | ET [h] | C v | C S.E. | $C_1$ v | $C_1$ S.E. | $C_2$ v | $C_2$ S.E. | $C_3$ v | $C_3$ S.E. | $C_4$ v | $C_4$ S.E. | $C_5$ v | $C_5$ S.E. | Sig. |
|---|---|---|---|---|---|---|---|---|---|---|---|---|---|---|
| 10 | 0 | 4,32 | 0,24 | 2,87 | 0,20 | 2,79 | 0,21 | 2,97 | 0,19 | 3,08 | 0,21 | 3,09 | 0,22 | |
| | 1 | 3,82 | 0,23 | 2,82 | 0,21 | 2,74 | 0,22 | 2,88 | 0,21 | 2,98 | 0,22 | 2,96 | 0,24 | |
| | 2 | 3,50 | 0,22 | 2,78 | 0,21 | 2,69 | 0,24 | 2,79 | 0,19 | 2,88 | 0,20 | 2,83 | 0,19 | |
| | 3 | 3,70 | 0,20 | 2,74 | 0,22 | 2,65 | 0,23 | 2,70 | 0,21 | 2,78 | 0,24 | 2,70 | 0,20 | |
| | 4 | 3,64 | 0,21 | 2,69 | 0,19 | 2,60 | 0,22 | 2,62 | 0,22 | 2,68 | 0,23 | 2,57 | 0,11 | |
| | 5 | 3,58 | 0,21 | 2,65 | 0,21 | 2,55 | 0,19 | 2,53 | 0,20 | 2,58 | 0,22 | 2,44 | 0,12 | |
| | 6 | 3,90 | 0,22 | 2,61 | 0,22 | 2,51 | 0,21 | 2,44 | 0,24 | 2,48 | 0,24 | 2,32 | 0,13 | |
| | 7 | 3,47 | 0,19 | 2,56 | 0,24 | 2,46 | 0,23 | 2,35 | 0,23 | 2,38 | 0,19 | 2,19 | 0,11 | |
| | 8 | 3,22 | 0,21 | 2,52 | 0,23 | 2,41 | 0,23 | 2,26 | 0,23 | 2,28 | 0,20 | 2,06 | 0,10 | significance level |
| | 9 | 3,26 | 0,48 | 2,58 | 0,13 | 2,37 | 0,14 | 2,17 | 0,20 | 2,18 | 0,19 | 1,78 | 0,30 | * P < 0,05 |
| | 10 | 3,29 | 0,24 | 2,43 | 0,19 | 2,32 | 0,20 | 2,08 | 0,22 | 2,07 | 0,12 | 1,80 | 0,10 | |
| | 11 | 3,23 | 0,23 | 2,39 | 0,24 | 2,27 | 0,21 | 1,99 | 0,23 | 1,97 | 0,11 | 1,67 | 0,10 | |
| | 12 | 3,18 | 0,23 | 2,34 | 0,23 | 2,23 | 0,21 | 1,91 | 0,19 | 1,87 | 0,13 | 1,55 | 0,10 | high significance level |
| | 13 | 3,13 | 0,50 | 2,30 | 0,20 | 2,18 | 0,21 | 1,82 | 0,46 | 1,65 | 0,14 | 1,42 | 0,09 | ** P < 0,01 |
| | 14 | 3,24 | 0,23 | 2,14 | 0,20 | 2,12 | 0,22 | 1,66 | 0,24 | 1,67 | 0,11 | 1,02 | 0,10 | |
| | 15 | 3,00 | 0,20 | 2,14 | 0,21 | 2,14 | 0,24 | 1,72 | 0,23 | 1,42 | 0,12 | 1,01 | 0,10 | |
| | 16 | 2,72 | 0,22 | 1,96 | 0,22 | 1,40 | 0,22 | 1,10 | 0,19 | 1,02 | 0,13 | 1,00 | 0,10 | |
| | 17 | 2,88 | 0,35 | 2,13 | 0,19 | 1,99 | 0,19 | 1,46 | 0,20 | 1,37 | 0,10 | 0,90 | 0,13 | |
| | 18 | 2,83 | 0,40 | 2,92 | 0,21 | 2,73 | 0,21 | 1,37 | 0,21 | 1,27 | 0,11 | 0,78 | 0,10 | |
| | 19 | 2,77 | 0,50 | 2,04 | 0,22 | 1,90 | 0,22 | 1,28 | 0,21 | 1,17 | 0,11 | 0,65 | 0,08 | |
| | 20 | 3,10 | 0,45 | 2,82 | 0,20 | 2,80 | 0,23 | 1,00 | 0,22 | 0,92 | 0,11 | 0,40 | 0,07 | |
| | 21 | 2,65 | 0,55 | 1,95 | 0,23 | 1,81 | 0,19 | 1,11 | 0,19 | 0,96 | 0,10 | 0,39 | 0,09 | |
| | 22 | 2,59 | 0,40 | 1,80 | 0,19 | 1,70 | 0,24 | 1,50 | 0,21 | 0,90 | 0,10 | 0,50 | 0,09 | |
| | 23 | 2,61 | 0,45 | 1,87 | 0,23 | 1,71 | 0,23 | 0,93 | 0,21 | 0,76 | 0,10 | 0,14 | 0,12 | |
| | 24 | 2,50 | 0,43 | 1,70 | 0,18 | 1,50 | 0,22 | 1,00 | 0,20 | 0,70 | 0,10 | 0,01 | 0,00 | |

**Tab. 28:** Average velocities [mm s$^{-1}$] of *D. magna* during different times of Cu-exposure.
Legend: N = Number of animals for each group; ET = Exposure Time; v = velocity [mm/s], S. E. = Standard Error of mean.

### Descriptives

v

| | N | Mean | Std. Deviation | Std. Error | 95% Confidence Interval for Mean Lower Bound | 95% Confidence Interval for Mean Upper Bound | Minimum | Maximum |
|---|---|---|---|---|---|---|---|---|
| C | 10 | 3,2620 | 1,5230 | ,4816 | 2,1725 | 4,3515 | 1,16 | 5,36 |
| C1 | 10 | 2,5800 | ,4201 | ,1328 | 2,2795 | 2,8805 | 1,94 | 3,02 |
| C2 | 10 | 2,3700 | ,4300 | ,1360 | 2,0624 | 2,6776 | 1,72 | 3,02 |
| C3 | 10 | 2,1700 | ,6201 | ,1961 | 1,7264 | 2,6136 | 1,22 | 3,12 |
| C4 | 10 | 2,1800 | ,6135 | ,1940 | 1,7411 | 2,6189 | 1,24 | 3,12 |
| C5 | 10 | 1,7800 | ,9576 | ,3028 | 1,0949 | 2,4651 | ,46 | 3,10 |
| Total | 60 | 2,3903 | ,9374 | ,1210 | 2,1482 | 2,6325 | ,46 | 5,36 |

**Tab. 29:** Descriptive Data for the parameter "swimming velocity" of *D. magna* after 9 h of Cu-exposure. N = Number of specimen.

Attachment

## One-way ANOVA

**Test of Homogeneity of Variances**

V

| Levene Statistic | df1 | df2 | Sig. |
|---|---|---|---|
| 3,811 | 5 | 54 | ,005 |

**ANOVA**

V

|  | Sum of Squares | df | Mean Square | F | Sig. |
|---|---|---|---|---|---|
| Between Groups | 12,615 | 5 | 2,523 | 3,473 | ,009 |
| Within Groups | 39,230 | 54 | ,726 | | |
| Total | 51,845 | 59 | | | |

V

Duncan[a]

| t | N | Subset for alpha = .05 | |
|---|---|---|---|
|   |   | 1 | 2 |
| C5 | 10 | 1,7800 | |
| C3 | 10 | 2,1700 | |
| C4 | 10 | 2,1800 | |
| C2 | 10 | 2,3700 | |
| C1 | 10 | 2,5800 | 2,5800 |
| C  | 10 |        | 3,2620 |
| Sig. | | ,065 | ,079 |

Means for groups in homogeneous subsets are displayed.
a. Uses Harmonic Mean Sample Size = 10,000.

**Tab. 30:** One-way ANOVA followed by Duncan´s multiple range post hoc test for the parameter "swimming velocity" of *D. magna* after 9 h of Cu-Exposure.

Attachment

## T-Test

**Group Statistics**

|   | t | N | Mean | Std. Deviation | Std. Error Mean |
|---|---|---|---|---|---|
| v | C | 10 | 3,2620 | 1,5230 | ,4816 |
|   | C1 | 10 | 2,5800 | ,4201 | ,1328 |

**Independent Samples Test**

|   |   | Levene's Test for Equality of Variances | | t-test for Equality of Means | | | | | | |
|---|---|---|---|---|---|---|---|---|---|---|
|   |   | F | Sig. | t | df | Sig. (2-tailed) | Mean Difference | Std. Error Difference | 95% Confidence Interval of the Difference | |
|   |   |   |   |   |   |   |   |   | Lower | Upper |
| v | Equal variances assumed | 7,415 | ,014 | 1,365 | 18 | ,189 | ,6820 | ,4996 | -,3676 | 1,7316 |
|   | Equal variances not assumed |   |   | 1,365 | 10,361 | ,201 | ,6820 | ,4996 | -,4259 | 1,7899 |

**Tab. 31:** t-test comparing the groups C and $C_1$ concerning the parameter "swimming velocity" of *D. magna* after 9 h of Cu-exposure. P > 0.05 not significant.

**Group Statistics**

|   | t | N | Mean | Std. Deviation | Std. Error Mean |
|---|---|---|---|---|---|
| v | C | 10 | 3,2620 | 1,5230 | ,4816 |
|   | C2 | 10 | 2,3700 | ,4300 | ,1360 |

**Independent Samples Test**

|   |   | Levene's Test for Equality of Variances | | t-test for Equality of Means | | | | | | |
|---|---|---|---|---|---|---|---|---|---|---|
|   |   | F | Sig. | t | df | Sig. (2-tailed) | Mean Difference | Std. Error Difference | 95% Confidence Interval of the Difference | |
|   |   |   |   |   |   |   |   |   | Lower | Upper |
| v | Equal variances assumed | 8,444 | ,009 | 1,782 | 18 | ,092 | ,8920 | ,5004 | -,1594 | 1,9434 |
|   | Equal variances not assumed |   |   | 1,782 | 10,426 | ,104 | ,8920 | ,5004 | -,2169 | 2,0009 |

**Tab. 32:** t-test comparing the groups C and $C_2$ concerning the parameter "swimming velocity" of *D. magna* after 9 h of Cu-exposure. P > 0.05 not significant.

Attachment

**Group Statistics**

| | t | N | Mean | Std. Deviation | Std. Error Mean |
|---|---|---|---|---|---|
| v | C | 10 | 3,2620 | 1,5230 | ,4816 |
| | $C_3$ | 10 | 2,1700 | ,6201 | ,1961 |

**Independent Samples Test**

| | | Levene's Test for Equality of Variances | | t-test for Equality of Means | | | | | | |
|---|---|---|---|---|---|---|---|---|---|---|
| | | F | Sig. | t | df | Sig. (2-tailed) | Mean Difference | Std. Error Difference | 95% Confidence Interval of the Difference | |
| | | | | | | | | | Lower | Upper |
| v | Equal variances assumed | 5,712 | ,028 | 2,100 | 18 | ,050 | 1,0920 | ,5200 | -4,97E-04 | 2,1845 |
| | Equal variances not assumed | | | 2,100 | 11,905 | ,058 | 1,0920 | ,5200 | -4,20E-02 | 2,2260 |

**Tab. 33:** t-test comparing the groups C and $C_3$ concerning the parameter "swimming velocity" of *D. magna* after 9 h of Cu-exposure. P > 0.05 not significant.

**Group Statistics**

| | t | N | Mean | Std. Deviation | Std. Error Mean |
|---|---|---|---|---|---|
| v | C | 10 | 3,2620 | 1,5230 | ,4816 |
| | $C_4$ | 10 | 2,1800 | ,6135 | ,1940 |

**Independent Samples Test**

| | | Levene's Test for Equality of Variances | | t-test for Equality of Means | | | | | | |
|---|---|---|---|---|---|---|---|---|---|---|
| | | F | Sig. | t | df | Sig. (2-tailed) | Mean Difference | Std. Error Difference | 95% Confidence Interval of the Difference | |
| | | | | | | | | | Lower | Upper |
| v | Equal variances assumed | 5,797 | ,027 | 2,084 | 18 | ,052 | 1,0820 | ,5192 | -8,83E-03 | 2,1728 |
| | Equal variances not assumed | | | 2,084 | 11,846 | ,060 | 1,0820 | ,5192 | -5,09E-02 | 2,2149 |

**Tab. 34:** : t-test comparing the groups C and $C_4$ concerning the parameter "swimming velocity" of *D. magna* after 9 h of Cu-exposure. P > 0.05 not significant.

Attachment

**Group Statistics**

|   |    | N  | Mean   | Std. Deviation | Std. Error Mean |
|---|----|----|--------|----------------|-----------------|
| v | C  | 10 | 3,2620 | 1,5230         | ,4816           |
|   | C5 | 10 | 1,7800 | ,9576          | ,3028           |

**Independent Samples Test**

| | | Levene's Test for Equality of Variances | | t-test for Equality of Means | | | | | | |
|---|---|---|---|---|---|---|---|---|---|---|
| | | F | Sig. | t | df | Sig. (2-tailed) | Mean Difference | Std. Error Difference | 95% Confidence Interval of the Difference | |
| | | | | | | | | | Lower | Upper |
| v | Equal variances assumed | 1,689 | ,210 | 2,605 | 18 | ,018 | 1,4820 | ,5689 | ,2868 | 2,6772 |
| | Equal variances not assumed | | | 2,605 | 15,155 | ,020 | 1,4820 | ,5689 | ,2705 | 2,6935 |

**Tab. 35:** t-test comparing the groups C and $C_5$ concerning the parameter "swimming velocity" of *D. magna* after 9 h of Cu-exposure. * $P < 0.05$ significant.

## Descriptive Data

**Descriptives**

v

|       | N  | Mean   | Std. Deviation | Std. Error | 95% Confidence Interval for Mean | | Minimum | Maximum |
|-------|----|--------|----------------|------------|-------|-------|---------|---------|
|       |    |        |                |            | Lower Bound | Upper Bound | | |
| C     | 10 | 3,1390 | 1,5864         | ,5017      | 2,0041 | 4,2739 | 1,12    | 5,14    |
| C1    | 10 | 2,3000 | ,6469          | ,2046      | 1,8372 | 2,7628 | 1,31    | 3,29    |
| C2    | 10 | 2,1800 | ,6663          | ,2107      | 1,7033 | 2,6567 | 1,17    | 3,19    |
| C3    | 10 | 1,8200 | 1,4500         | ,4585      | ,7827  | 2,8573 | ,03     | 3,61    |
| C4    | 10 | 1,6500 | ,4517          | ,1428      | 1,3269 | 1,9731 | ,78     | 2,52    |
| C5    | 10 | 1,4190 | ,2769          | 8,757E-02  | 1,2209 | 1,6171 | 1,00    | 1,84    |
| Total | 60 | 2,0847 | 1,0935         | ,1412      | 1,8022 | 2,3672 | ,03     | 5,14    |

**Tab. 36:** Descriptive Data for the parameter "swimming velocity" of *D. magna* after 13 h of Cu-exposure. N = Number of specimen.

Attachment

## One-way ANOVA

**Test of Homogeneity of Variances**

v

| Levene Statistic | df1 | df2 | Sig. |
|---|---|---|---|
| 13,176 | 5 | 54 | ,000 |

**ANOVA**

v

|  | Sum of Squares | df | Mean Square | F | Sig. |
|---|---|---|---|---|---|
| Between Groups | 18,692 | 5 | 3,738 | 3,892 | ,004 |
| Within Groups | 51,862 | 54 | ,960 |  |  |
| Total | 70,554 | 59 |  |  |  |

v

Duncan[a]

| t | N | Subset for alpha = .05 | |
|---|---|---|---|
|  |  | 1 | 2 |
| C5 | 10 | 1,4190 |  |
| C4 | 10 | 1,6500 |  |
| C3 | 10 | 1,8200 |  |
| C2 | 10 | 2,1800 |  |
| C1 | 10 | 2,3000 | 2,3000 |
| C | 10 |  | 3,1390 |
| Sig. |  | ,077 | ,061 |

Means for groups in homogeneous subsets are displayed.
a. Uses Harmonic Mean Sample Size = 10,000.

**Tab. 37:** One-way ANOVA followed by Duncan´s multiple range post hoc test for the parameter "swimming velocity" of *D. magna* after 13 h of Cu-Exposure.

## T-Test

**Group Statistics**

|  | t | N | Mean | Std. Deviation | Std. Error Mean |
|---|---|---|---|---|---|
| v | C | 10 | 3,1390 | 1,5864 | ,5017 |
|  | C1 | 10 | 2,3000 | ,6469 | ,2046 |

Attachment

**Independent Samples Test**

| | | Levene's Test for Equality of Variances | | t-test for Equality of Means | | | | | | |
|---|---|---|---|---|---|---|---|---|---|---|
| | | | | | | | | | 95% Confidence Interval of the Difference | |
| | | F | Sig. | t | df | Sig. (2-tailed) | Mean Difference | Std. Error Difference | Lower | Upper |
| v | Equal variances assumed | 10,312 | ,005 | 1,549 | 18 | ,139 | ,8390 | ,5418 | -,2992 | 1,9772 |
| | Equal variances not assumed | | | 1,549 | 11,912 | ,148 | ,8390 | ,5418 | -,3424 | 2,0204 |

**Tab. 38:** t-test comparing the groups C and $C_1$ concerning the parameter "swimming velocity" of *D. magna* after 13 h of Cu-exposure. P > 0.05 not significant.

**Group Statistics**

| | t | N | Mean | Std. Deviation | Std. Error Mean |
|---|---|---|---|---|---|
| v | C | 10 | 3,1390 | 1,5864 | ,5017 |
| | C2 | 10 | 2,1800 | ,6663 | ,2107 |

**Independent Samples Test**

| | | Levene's Test for Equality of Variances | | t-test for Equality of Means | | | | | | |
|---|---|---|---|---|---|---|---|---|---|---|
| | | | | | | | | | 95% Confidence Interval of the Difference | |
| | | F | Sig. | t | df | Sig. (2-tailed) | Mean Difference | Std. Error Difference | Lower | Upper |
| v | Equal variances assumed | 9,487 | ,006 | 1,762 | 18 | ,095 | ,9590 | ,5441 | -,1842 | 2,1022 |
| | Equal variances not assumed | | | 1,762 | 12,080 | ,103 | ,9590 | ,5441 | -,2257 | 2,1437 |

**Tab. 39:** t-test comparing the groups C and $C_2$ concerning the parameter "swimming velocity" of *D. magna* after 13 h of Cu-exposure. P > 0.05 not significant.

## Attachment

**Group Statistics**

| | t | N | Mean | Std. Deviation | Std. Error Mean |
|---|---|---|---|---|---|
| v | C | 10 | 3,1390 | 1,5864 | ,5017 |
| | C3 | 10 | 1,8200 | 1,4500 | ,4585 |

**Independent Samples Test**

| | | Levene's Test for Equality of Variances | | t-test for Equality of Means | | | | | 95% Confidence Interval of the Difference | |
|---|---|---|---|---|---|---|---|---|---|---|
| | | F | Sig. | t | df | Sig. (2-tailed) | Mean Difference | Std. Error Difference | Lower | Upper |
| v | Equal variances assumed | ,017 | ,899 | 1,941 | 18 | ,068 | 1,3190 | ,6797 | -,1089 | 2,7469 |
| | Equal variances not assumed | | | 1,941 | 17,856 | ,068 | 1,3190 | ,6797 | -,1097 | 2,7477 |

**Tab. 40:** t-test comparing the groups C and $C_3$ concerning the parameter "swimming velocity" of *D. magna* after 13 h of Cu-exposure. P > 0.05 not significant.

**Group Statistics**

| | t | N | Mean | Std. Deviation | Std. Error Mean |
|---|---|---|---|---|---|
| v | C | 10 | 3,1390 | 1,5864 | ,5017 |
| | C4 | 10 | 1,6500 | ,4517 | ,1428 |

**Independent Samples Test**

| | | Levene's Test for Equality of Variances | | t-test for Equality of Means | | | | | 95% Confidence Interval of the Difference | |
|---|---|---|---|---|---|---|---|---|---|---|
| | | F | Sig. | t | df | Sig. (2-tailed) | Mean Difference | Std. Error Difference | Lower | Upper |
| v | Equal variances assumed | 15,669 | ,001 | 2,855 | 18 | ,011 | 1,4890 | ,5216 | ,3931 | 2,5849 |
| | Equal variances not assumed | | | 2,855 | 10,450 | ,016 | 1,4890 | ,5216 | ,3335 | 2,6445 |

**Tab. 41:** t-test comparing the groups C and $C_4$ concerning the parameter "swimming velocity" of *D. magna* after 13 h of Cu-exposure. * P < 0.05 significant.

## Group Statistics

| | t | N | Mean | Std. Deviation | Std. Error Mean |
|---|---|---|---|---|---|
| v | C | 10 | 3,1390 | 1,5864 | ,5017 |
| | C5 | 10 | 1,4190 | ,2769 | 8,757E-02 |

## Independent Samples Test

| | | Levene's Test for Equality of Variances | | t-test for Equality of Means | | | | | | |
|---|---|---|---|---|---|---|---|---|---|---|
| | | F | Sig. | t | df | Sig. (2-tailed) | Mean Difference | Std. Error Difference | 95% Confidence Interval of the Difference | |
| | | | | | | | | | Lower | Upper |
| v | Equal variances assumed | 20,941 | ,000 | 3,377 | 18 | ,003 | 1,7200 | ,5093 | ,6501 | 2,7899 |
| | Equal variances not assumed | | | 3,377 | 9,548 | ,008 | 1,7200 | ,5093 | ,5780 | 2,8620 |

**Tab. 42:** t-test comparing the groups C and $C_5$ concerning the parameter "swimming velocity" of *D. magna* after 13 h of Cu-exposure. ** $P < 0.01$ high significant.

## Descriptive Data

### Descriptives

v

| | N | Mean | Std. Deviation | Std. Error | 95% Confidence Interval for Mean | | Minimum | Maximum |
|---|---|---|---|---|---|---|---|---|
| | | | | | Lower Bound | Upper Bound | | |
| C | 10 | 3,0700 | 1,4647 | ,4632 | 2,0222 | 4,1178 | ,96 | 5,18 |
| C1 | 10 | 2,3410 | 1,1632 | ,3679 | 1,5089 | 3,1731 | ,24 | 4,04 |
| C2 | 10 | 2,1200 | 1,3884 | ,4390 | 1,1268 | 3,1132 | ,13 | 4,15 |
| C3 | 10 | 1,6600 | ,9945 | ,3145 | ,9486 | 2,3714 | ,64 | 3,68 |
| C4 | 10 | 1,6700 | ,3036 | 9,600E-02 | 1,4528 | 1,8872 | 1,05 | 2,29 |
| C5 | 10 | 1,0180 | ,2941 | 9,301E-02 | ,8076 | 1,2284 | ,47 | 1,57 |
| Total | 60 | 1,9798 | 1,1929 | ,1540 | 1,6717 | 2,2880 | ,13 | 5,18 |

**Fig. 28:** Descriptive Data for the parameter "swimming velocity" of *D. magna* after 14 h of Cu-exposure. N = Number of specimen.

Attachment

## One-way ANOVA

**Test of Homogeneity of Variances**

v

| Levene Statistic | df1 | df2 | Sig. |
|---|---|---|---|
| 4,549 | 5 | 54 | ,002 |

**ANOVA**

v

| | Sum of Squares | df | Mean Square | F | Sig. |
|---|---|---|---|---|---|
| Between Groups | 24,620 | 5 | 4,924 | 4,481 | ,002 |
| Within Groups | 59,344 | 54 | 1,099 | | |
| Total | 83,964 | 59 | | | |

v

Duncan[a]

| t | N | Subset for alpha = .05 | | |
|---|---|---|---|---|
| | | 1 | 2 | 3 |
| C5 | 10 | 1,0180 | | |
| C3 | 10 | 1,6600 | 1,6600 | |
| C4 | 10 | 1,6700 | 1,6700 | |
| C2 | 10 | | 2,1200 | 2,1200 |
| C1 | 10 | | 2,3410 | 2,3410 |
| C | 10 | | | 3,0700 |
| Sig. | | ,196 | ,192 | ,060 |

Means for groups in homogeneous subsets are displayed.
a. Uses Harmonic Mean Sample Size = 10,000.

**Fig. 29:** One-way ANOVA followed by Duncan´s multiple range post hoc test for the parameter "swimming velocity" of *D. magna* after 14 h of Cu-Exposure.

# Attachment

## T-Test

**Group Statistics**

|   | t | N | Mean | Std. Deviation | Std. Error Mean |
|---|---|---|---|---|---|
| v | C | 10 | 3,0700 | 1,4647 | ,4632 |
|   | C1 | 10 | 2,3410 | 1,1632 | ,3679 |

**Independent Samples Test**

| | | Levene's Test for Equality of Variances | | t-test for Equality of Means | | | | | | |
|---|---|---|---|---|---|---|---|---|---|---|
| | | F | Sig. | t | df | Sig. (2-tailed) | Mean Difference | Std. Error Difference | 95% Confidence Interval of the Difference | |
| | | | | | | | | | Lower | Upper |
| v | Equal variances assumed | ,623 | ,440 | 1,232 | 18 | ,234 | ,7290 | ,5915 | -,5137 | 1,9717 |
| | Equal variances not assumed | | | 1,232 | 17,122 | ,234 | ,7290 | ,5915 | -,5183 | 1,9763 |

**Tab. 43**: t-test comparing the groups C and $C_1$ concerning the parameter "swimming velocity" of *D. magna* after 13 h of Cu-exposure. P > 0.05 not significant.

**Group Statistics**

|   | t | N | Mean | Std. Deviation | Std. Error Mean |
|---|---|---|---|---|---|
| v | C | 10 | 3,0700 | 1,4647 | ,4632 |
|   | C2 | 10 | 2,1200 | 1,3884 | ,4390 |

**Independent Samples Test**

| | | Levene's Test for Equality of Variances | | t-test for Equality of Means | | | | | | |
|---|---|---|---|---|---|---|---|---|---|---|
| | | F | Sig. | t | df | Sig. (2-tailed) | Mean Difference | Std. Error Difference | 95% Confidence Interval of the Difference | |
| | | | | | | | | | Lower | Upper |
| v | Equal variances assumed | ,229 | ,638 | 1,489 | 18 | ,154 | ,9500 | ,6382 | -,3908 | 2,2908 |
| | Equal variances not assumed | | | 1,489 | 17,949 | ,154 | ,9500 | ,6382 | -,3911 | 2,2911 |

**Tab. 44:** t-test comparing the groups C and $C_2$ concerning the parameter "swimming velocity" of *D. magna* after 14 h of Cu-exposure. P > 0.05 not significant.

## Attachment

**Group Statistics**

| | t | N | Mean | Std. Deviation | Std. Error Mean |
|---|---|---|---|---|---|
| v | C | 10 | 3,0700 | 1,4647 | ,4632 |
| | C3 | 10 | 1,6600 | ,9945 | ,3145 |

**Independent Samples Test**

| | | Levene's Test for Equality of Variances | | t-test for Equality of Means | | | | | |
|---|---|---|---|---|---|---|---|---|---|
| | | | | | | | | 95% Confidence Interval of the Difference | |
| | | F | Sig. | t | df | Sig. (2-tailed) | Mean Difference | Std. Error Difference | Lower | Upper |
| v | Equal variances assumed | 1,682 | ,211 | 2,518 | 18 | ,021 | 1,4100 | ,5599 | ,2338 | 2,5862 |
| | Equal variances not assumed | | | 2,518 | 15,843 | ,023 | 1,4100 | ,5599 | ,2222 | 2,5978 |

**Tab. 45:** t-test comparing the groups C and $C_3$ concerning the parameter "swimming velocity" of *D. magna* after 14 h of Cu-exposure. * P < 0.05 significant.

**Group Statistics**

| | t | N | Mean | Std. Deviation | Std. Error Mean |
|---|---|---|---|---|---|
| v | C | 10 | 3,0700 | 1,4647 | ,4632 |
| | C4 | 10 | 1,6700 | ,3036 | 9,600E-02 |

**Independent Samples Test**

| | | Levene's Test for Equality of Variances | | t-test for Equality of Means | | | | | |
|---|---|---|---|---|---|---|---|---|---|
| | | | | | | | | 95% Confidence Interval of the Difference | |
| | | F | Sig. | t | df | Sig. (2-tailed) | Mean Difference | Std. Error Difference | Lower | Upper |
| v | Equal variances assumed | 14,441 | ,001 | 2,960 | 18 | ,008 | 1,4000 | ,4730 | ,4062 | 2,3938 |
| | Equal variances not assumed | | | 2,960 | 9,772 | ,015 | 1,4000 | ,4730 | ,3427 | 2,4573 |

**Tab. 46:** t-test comparing the groups C and $C_4$ concerning the parameter "swimming velocity" of *D. magna* after 14 h of Cu-exposure. * P < 0.05 significant.

Attachment

**Group Statistics**

|   | t | N | Mean | Std. Deviation | Std. Error Mean |
|---|---|---|---|---|---|
| v | C | 10 | 3,0700 | 1,4647 | ,4632 |
|   | C5 | 10 | 1,0180 | ,2941 | 9,301E-02 |

**Independent Samples Test**

| | | Levene's Test for Equality of Variances | | t-test for Equality of Means | | | | | | |
|---|---|---|---|---|---|---|---|---|---|---|
| | | | | | | | | | 95% Confidence Interval of the Difference | |
| | | F | Sig. | t | df | Sig. (2-tailed) | Mean Difference | Std. Error Difference | Lower | Upper |
| v | Equal variances assumed | 14,269 | ,001 | 4,343 | 18 | ,000 | 2,0520 | ,4724 | 1,0594 | 3,0446 |
|   | Equal variances not assumed | | | 4,343 | 9,725 | ,002 | 2,0520 | ,4724 | ,9953 | 3,1087 |

**Tab. 47:** t-test comparing the groups C and $C_5$ concerning the parameter "swimming velocity" of *D. magna* after 14 h of Cu-exposure. ** $P < 0.01$ high significant.

### VII. 2. 1. 3. *Daphnia pulex* LEYDIG

Variable:  v

Gruppiert nach:  t

|   | Quadrat-summe | F.G. | mittlere QS | F | P |
|---|---|---|---|---|---|
| Zwischen | 4,83646 | 2 | 2,41823 | 15,8004 | 7,65138e-05 |
| Innerhalb | 3,06098 | 20 | 0,15304 | | |
| Gesamt | 7,89744 | 22 | 0,35897 | | |

**Bartlett-Test zur Varianzengleichheit**

| Chi-Quadrat | F.G. | P |
|---|---|---|
| 6,14719 | 2 | 0,04625 |

Attachment

**Multiple Vergleiche nach Methode: Duncan**

Signifikanzniveau: 0,05

| Anzahl Gruppen in Untermenge | Kritischer Wert R für signifikanten Unterschied |
|---|---|
| 2 | 2,95001 |
| 3 | 3,09753 |

Signifikante Unterschiede zwischen Gruppenpaaren sind in folgender Tabelle mit '*' gekennzeichnet:

**Mittelwert**

| Mittelwert | C1 | C2 | C |
|---|---|---|---|
| 0,25285 | | - | + |
| 0,51250 | | | + |
| 1,32875 | | * | * |

**Tab. 48:** One-way ANOVA followed by Duncan's multiple range post hoc test for the parameter "swimming velocity" of *D. pulex* after 9 h of Cu-Exposure.

**Variable:** v

**Gruppiert nach:** t

|  | N | Mittelwert | Std.Abw. |
|---|---|---|---|
| Gesamte Stichprobe | 20 | 0,82666 | 0,61412 |
| C | 10 | 1,32875 | 0,19845 |
| C1 | 10 | 0,25285 | 0,33737 |

| F | P | Varianz-schätzung | T | F.G. | P |
|---|---|---|---|---|---|
| 2,89011 | 0,19107 | Homogen | 7,65532 | 18 | 3,60945e-06 |
| | | Heterogen | 7,39213 | 9,44128 | 3,19827e-05 |

**Tab. 49:** t-test comparing the groups C and $C_1$ concerning the parameter "swimming velocity" of *D. pulex* after 9 h of Cu-exposure. *** $P < 0.001$ very high significant.

Attachment

Variable: v

Gruppiert nach: t

| | N | Mittelwert | Std.Abw. |
|---|---|---|---|
| Gesamte Stichprobe | 20 | 0,92062 | 0,57983 |
| C | 10 | 1,32875 | 0,19845 |
| C2 | 10 | 0,51250 | 0,54802 |

| F | P | Varianz-schätzung | T | F.G. | P |
|---|---|---|---|---|---|
| 7,62585 | 0,01561 | Homogen | 3,96103 | 18 | 0,00142 |
| | | Heterogen | 3,96103 | 8,80483 | 0,00344 |

**Tab. 50:** t-test comparing the groups C and $C_2$ concerning the parameter "swimming velocity" of *D. pulex* after 9 h of Cu-exposure. ** P < 0.01 high significant.

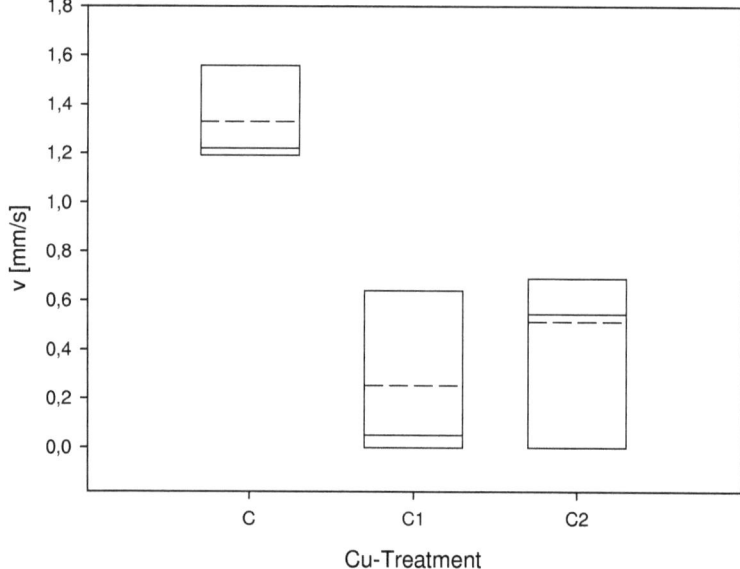

**Fig. 30:** Effects of Cu on swimming velocity of D. pulex after 9 h of permanent Cu-Exposure. Data are depicted as Median (solid line), Mean (dashed line), $25^{th}$ and $75^{th}$ percentile (boundary of boxes)

Attachment

Variable: v

Gruppiert nach: t

| | Quadrat-summe | F.G. | mittlere QS | F | P |
|---|---|---|---|---|---|
| Zwischen | 2,64992 | 2 | 1,32496 | 6,37428 | 0,00686 |
| Innerhalb | 4,36508 | 21 | 0,20786 | | |
| Gesamt | 7,01500 | 23 | 0,30500 | | |

Bartlett-Test zur Varianzengleichheit

| Chi-Quadrat | F.G. | P |
|---|---|---|
| 0,12602 | 2 | 0,93893 |

Multiple Vergleiche nach Methode: Duncan

Signifikanzniveau: 0,05

| Anzahl Gruppen in Untermenge | Kritischer Wert R für signifikanten Unterschied |
|---|---|
| 2 | 2,94104 |
| 3 | 3,08838 |

Obige Werte sind Tabellenwerte 'R'. Die tatsächlichen Vergleichswerte berechnen sich jeweils aus der Formel:

0,32238 * R * SQRT( 1/N1 + 1/N2 )

Signifikante Unterschiede zwischen Gruppenpaaren sind in folgender Tabelle mit '*' gekennzeichnet:

Mittelwert

| | | |
|---|---|---|
| 0,52375 | C2 | - + |
| 0,64125 | C1 | + |
| 1,28000 | C | * * |

Untermenge 1: C2
              C1

Untermenge 2: C

Fig. 31: One-way ANOVA followed by Duncan´s multiple range post hoc test for the parameter "swimming velocity" of *D. pulex* after 11 h of Cu-Exposure.

Attachment

Variable: v

Gruppiert nach: t

|  | N | Mittelwert | Std.Abw. |
|---|---|---|---|
| Gesamte Stichprobe | 20 | 0,96062 | 0,56355 |
| C | 10 | 1,28000 | 0,46687 |
| C1 | 10 | 0,64125 | 0,47900 |

| F | P | Varianz-schätzung | T | F.G. | P |
|---|---|---|---|---|---|
| 1,05262 | 0,9478 | Homogen | 2,70098 | 18 | 0,01722 |
|  |  | Heterogen | 2,70098 | 13,9908 | 0,01723 |

**Fig. 32:** t-test comparing the groups C and $C_1$ concerning the parameter "swimming velocity" of *D. pulex* after 11 h of Cu-exposure. * $P < 0.05$ significant.

Variable: v

Gruppiert nach: t

|  | N | Mittelwert | Std.Abw. |
|---|---|---|---|
| Gesamte Stichprobe | 20 | 0,90187 | 0,58003 |
| C | 10 | 1,28000 | 0,46687 |
| C2 | 10 | 0,52375 | 0,41972 |

| F | P | Varianz-schätzung | T | F.G. | P |
|---|---|---|---|---|---|
| 1,23728 | 0,78595 | Homogen | 3,40710 | 18 | 0,00425 |
|  |  | Heterogen | 3,40710 | 13,8443 | 0,00431 |

**Fig. 33:** t-test comparing the groups C and $C_2$ concerning the parameter "swimming velocity" of *D. pulex* after 11 h of Cu-exposure. ** $P < 0.01$ high significant.

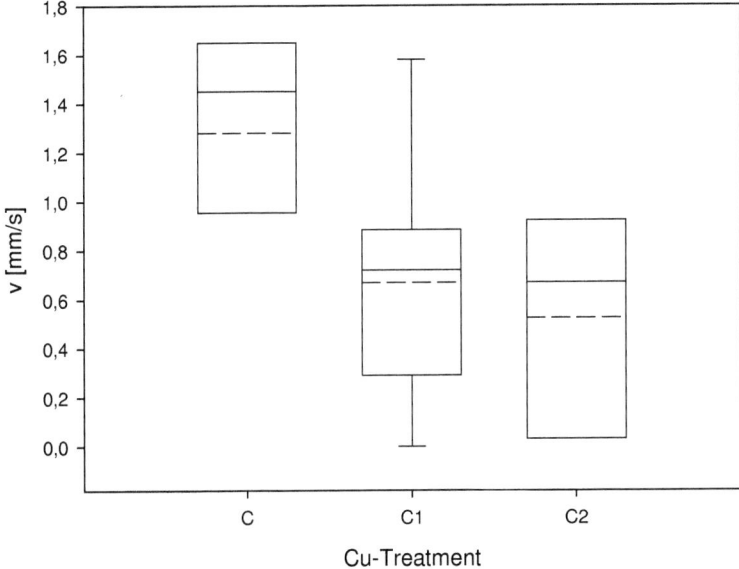

**Fig. 34:** Effects of Cu on swimming velocity of *D. pulex* after 9 h of permanent Cu-Exposure. Data are depicted as Median (solid line), Mean (dashed line), $25^{th}$ and $75^{th}$ percentile (boundary of boxes). $10^{th}$ and $90^{th}$ percentile (whiskers).

## Summary

In this study changes in the locomotory behaviour of three different crustacean species (*D. magna* Straus, *D. pulex* Leydig, *H. inermis* Leach) have been used for studying sublethal effects of the heavy metals Cu and Cd. The behavioural responses were determined by a real time image analysis, using a video camera and a Pentium-PC equipped with a standard low cost frame grabber. For a sequence of 50 images per group, where 10 test organisms were moving simultaneously, the trajectories have been reconstructed in binary image sequences. As biological end points we defined the average swimming velocity and the average duration of swimming activity and inactivity. The behavioural responses of the crustaceans were analysed under normal conditions (without heavy metal stress) and after application of sublethal concentration of heavy metal. The sublethal concentrations are based on determined $LC_{50}$ values. In the trial with *D. magna* Cu [in $\mu g\ l^{-1}$ (ppb)]: $C_1 = 1$, $C_2 = 5$, $C_3 = 10$, $C_4 = 20$, $C_5 = 30$ have been used. In the trial with *D. pulex* following Cu concentrations have been used: $C_1 = 5$ ppb, $C_2 = 10$ ppb. The test organisms were exposed to the Cu concentration for 24 hours under static conditions. Already after 9 hours of Cu-exposure to *D. magna* a significant (* $P < 0.05$) decrease of the average swimming velocity could be observed at the group of the highest Cu concentration (30 ppb). After 13 hours of Cu-contamination the swimming velocity was significantly (* $P < 0.05$) reduced at group of 20 ppb Cu-treatment and after 14 hours a significant (* $P < 0.05$) decrease of the average swimming velocity could be measured at the group of 10 ppb Cu-treatment. No significant decrease of the swimming velocity could be observed in the 1 ppb and 5 ppb Cu-treatment. Already after 9 hours of Cu-exposure to *D. pulex* test animals showed a very high significant (*** $P < 0.001$; ) decrease of the average swimming velocity at $C_1$ and a high significant (** $P < 0.01$) decrease at $C_2$.

The marine crustacean *Hippolyte inermis* Leach have been used to study acute and sublethal effects of the non essential heavy metal cadmium. Test animals were stressed by Cd of following concentrations: $C_1 = 1$ ppm, $C_2 = 2$ ppm, $C_3 = 3.5$ ppm. The shrimps were exposed to the heavy metal concentrations for 12 hours under static conditions. Already at the start time (0 h) of Cd-exposure test animals showed a high significant (** $P < 0.01$; ) decrease of the average swimming velocity at $C_3$. After 1 h of Cd exposure, median moving velocity significantly decreased at $C_2$ (* $P \leq 0.05$). After 3 hours of Cd exposure, median moving velocity was for the first time high significantly reduced in the 1 ppm Cd-treatment (** $P \leq 0.01$).

Summary

## Zusammenfassung

Mit vorliegender Studie wurde das lokomotorische Verhalten von drei Crustaceenarten (*D. magna* Straus, *D. pulex* LEYDIG, *H. inermis* LEACH) herangezogen, um subletale Effekte von den Schwermetallen Kupfer und Cadmium zu untersuchen. Die Verhaltensreaktionen wurden mittels Echtzeit-Videoanalyse ausgewertet. Für eine Sequenz von 50 Bildern pro Gruppe, in welcher sich 10 Organismen simultan bewegten, wurden die Trajektorien rekonstruiert in digitalen Bildsequenzen. Als biologische Endpunkte wurden die Parameter „Schwimmgeschwindigkeit" und „Schwimmaktivität" definiert. Das Stressverhalten wurde dabei unter normalen Bedingungen (ohne Schwermetallstress) und unter subletalen Konzentrationen analysiert. Die subletalen Konzentrationen basieren auf bekannten $LC_{50}$-Werten. Im Cu-Versuch mit *D. magna* wurden folgende Cu-lösungen verwendet [in $\mu g\ l^{-1}$ (ppb)]: $C_1 = 1$, $C_2 = 5$, $C_3 = 10$, $C_4 = 20$, $C_5 = 30$. Im Cu-Versuch mit *D. pulex* wurden folgende Cu-lösungen verwendet in [$\mu g\ l^{-1}$ (ppb)]: $C_1 = 5$ ppb, $C_2 = 10$. Bereits nach 9 Stunden Cu Exposition zeigte *D. magna* eine signifikante (* $P < 0.015$) Verringerung der Schwimmgeschwindigkeit in der Gruppe mit der höchsten Kupferkonzentration (30 ppb). Nach 13 Stunden permanenter Kupferexposition wurde die Schwimmgeschwindigkeit signifikant (* $P < 0.05$) in der 20 ppb Gruppe reduziert und nach 14 h konnte eine signifikante Verringerung in der 10 ppb Gruppe nachgewiesen werden. Auch bei *D. pulex* konnten bereits nach 9 h hoch signifikante Verringerungen (*** $P < 0.001$; ) der Schwimmgeschwindigkeit in der 5 ppb Gruppe und in der 10 ppb (** $P < 0.01$) festgestellt werden.

Der marine Crustacee *Hippolyte inermis* LEACH wurde verwendet, um akute und subletale Effecte des nicht essentiellen Schwermetalls Cadmium zu untersuchen. Die Testorganismen wurden mit folgenden Cd-Konzentrationen gestresst: $C = 0$, $C_1 = 1$ ppm, $C_2 = 2$ ppm, $C_3 = 3.5$ ppm. Die Seegrasgarnelen wurden für 12 Stunden den Metallkonzentrationen unter statischen Bedingungen, d. h. ohne Wasseraustausch, ausgesetzt. Bereits zu Beginn des Versuches zeigten die Tiere eine hoch signifikante (** $P < 0,01$) Verringerung der Schwimm- bzw. Bewegungsaktivität. Nach 1 h permanenter Cd Exposition wurde die mittlere Bewegungsgeschwindigkeit (gemessen am Median) in der 2 ppm Gruppe signifikant verringert. Nach 3 Stunden wurde die Bewegungsaktivität in der 1 ppm Gruppe signifikant verringert.

VDM Verlagsservicegesellschaft mbH

Die VDM Verlagsservicegesellschaft sucht für wissenschaftliche Verlage abgeschlossene und herausragende

## Dissertationen, Habilitationen, Diplomarbeiten, Master Theses, Magisterarbeiten usw.

für die kostenlose Publikation als Fachbuch.

Sie verfügen über eine Arbeit, die hohen inhaltlichen und formalen Ansprüchen genügt, und haben Interesse an einer honorarvergüteten Publikation?

Dann senden Sie bitte erste Informationen über sich und Ihre Arbeit per Email an *info@vdm-vsg.de*.

**Sie erhalten kurzfristig unser Feedback!**

VDM Verlagsservicegesellschaft mbH
Dudweiler Landstr. 99           Telefon  +49 681 3720 174
D - 66123 Saarbrücken           Fax      +49 681 3720 1749

**www.vdm-vsg.de**

Die VDM Verlagsservicegesellschaft mbH vertritt

Printed by Books on Demand GmbH, Norderstedt / Germany